Heteromation, and Other Stories of Computing and Capitalism

Acting with Technology

Bonnie Nardi, Victor Kaptelinin, and Kirsten Foot, editors

Tracing Genres through Organizations: A Sociocultural Approach to Information Design, Clay Spinuzzi, 2003

Activity-Centered Design: An Ecological Approach to Designing Smart Tools and Usable Systems, Geri Gay and Helene Hembrooke, 2004

The Semiotic Engineering of Human Computer Interaction, Clarisse Sieckenius de Souza, 2005

Group Cognition: Computer Support for Building Collaborative Knowledge, Gerry Stahl, 2006

Acting with Technology: Activity Theory and Interaction Design, Victor Kaptelinin and Bonnie A. Nardi, 2006

Web Campaigning, Kirsten A. Foot and Steven M. Schneider, 2006

Scientific Collaboration on the Internet, Gary M. Olson, Ann Zimmerman, and Nathan Bos, editors, 2008

Acting with Technology: Activity Theory and Interaction Design, Victor Kaptelinin and Bonnie A. Nardi, 2009

Digitally-Enabled Social Change: Online and Offline Activism in the Age of the Internet, Jennifer Earl and Katrina Kimport, 2011

Invisible Users: Youth in the Internet Cafés of Urban Ghana, Jenna Burrell, 2012

Venture Labor: Work and the Burden of Risk in Innovative Industries, Gina Neff, 2012

Car Crashes without Cars: Lessons about Simulation Technology and Organizational Change from Automotive Design, Paul M. Leonardi, 2012

Coding Places: Software Practice in a South American City, Yuri Takhteyev, 2012

Technology Choices: Why Occupations Differ in Their Embrace of New Technology, Diane E. Bailey and Paul M. Leonardi, 2015

Shifting Practices: A Reflective Inquiry into Technology, Practice, and Innovation, Giovan Francesco Lanzara, 2016

Heteromation, and Other Stories of Computing and Capitalism, Hamid R. Ekbia and Bonnie Nardi, 2017

Heteromation, and Other Stories of Computing and Capitalism

Hamid R. Ekbia and Bonnie Nardi

The MIT Press
Cambridge, Massachusetts
London, England

© 2017 Massachusetts Institute of Technology

All rights reserved. No part of this book may be reproduced in any form by any electronic or mechanical means (including photocopying, recording, or information storage and retrieval) without permission in writing from the publisher.

This book was set in Sabon LT Std by Toppan Best-set Premedia Limited. Printed and bound in the United States of America.

Library of Congress Cataloging-in-Publication Data

Names: Ekbia, H. R. (Hamid Reza), 1955- author. | Nardi, Bonnie A., author.
Title: Heteromation, and other stories of computing and capitalism / Hamid R. Ekbia and Bonnie A. Nardi.
Description: Cambridge, MA : MIT Press, [2017] | Series: Acting with technology | Includes bibliographical references and index.
Identifiers: LCCN 2016043785 | ISBN 9780262036252 (hardcover : alk. paper)
Subjects: LCSH: Labor theory of value. | Technological innovation--Economic aspects. | Capitalism.
Classification: LCC HB206 .E 2017 | DDC 331.2/5--dc23 LC record available at https://lccn.loc.gov/2016043785

10 9 8 7 6 5 4 3 2 1

To our children:
Kaveh, Kia, and Taraneh
Anthony, Christopher, and Jeanette
for the challenges and excitements they face.

Contents

Preface ix

Prologue: An (Untold) Story of Economy and Computing 1

I Looking Back

1 Heteromation: A Revised (His)story of Computing 23
2 The Logic of Wealth Accumulation: A Story of Political Economy 41
3 The Dynamics of Capitalist Change: A Story of Resilience 57
4 Possibilities and Predicaments: A Story of Stimulus 81

II Varieties of Heteromated Labor

5 Communicative Labor: A Story of Separation 93
6 Cognitive Labor: A Story of Mental Toil 107
7 Creative Labor: A Story of Mental Magic 117
8 Emotional Labor: A Story of Caring 129
9 Organizing Labor: A Story of Commitment 147

III Looking Ahead

10 Mechanisms of Participation: A Story of Rewards (and Punishments) 161
11 The Governance of Social Life: A Story of "Work" and Institutional Implosion 177

12 Utopias: A Story of Revolution and Reform   187
Epilogue: The Story of Machines and Us   211

Notes   225
References   235
Index   257

Preface

Academic writing is a delicate and tricky undertaking. The standards that govern this form of writing are suboptimal at best, self-defeating otherwise. The reason is that the standards tend to tie analysis to very specific methodologies applied at a particular "level." This level-specific approach is at the foundation of how academic disciplines are traditionally organized, and while it might work for some topics some of the time, it does not for others most of the time. A case in point is the topic of the present book, which seeks to explore the relationship between computing and capitalism, with a focus on the division of labor between humans and machines. This relationship can be traced on many different levels, layers, and rhythms, from individual psychology and subjectivity all the way to the economic, political, and epochal changes of the embedding social system. As such, the topic can be meaningfully studied only through a conceptual lens that can smoothly shift back and forth between different levels of analysis. This is the tack that we have followed in this writing.

Fortunately for us, there are outstanding examples of such writing, most prominently by thinkers and writers whose ideas have deeply influenced us in other ways: Karl Marx, Michel Foucault, Max Weber, to name a few. To "apply" the method of these thinkers to the circumstances that we seek to understand, however, we found ourselves in need of conceptual innovations, of which the term "heteromation" is a key example. We coined and introduced the term, knowing full well the conceptual risks and intellectual pitfalls of such innovation. In particular, we were concerned about two things: (1) the risk of conceptual vacuity—that is, of a concept with little empirical support and grounding; and (2) the pitfall of redundancy, reinventing the wheel, or what one astute reviewer of our manuscript described as the common phenomenon of "term entrepreneurship."

To deal with the first concern, we have tried to ground our concepts in a wide range of empirical facts and observations, which are mainly provided in part II of the book. In regard to the second pitfall, we are quite sympathetic to the reviewer's self-reflexive observation that finds in much of academic writing "an extension of a capitalist logic not unlike that seen in commodity fetish and trademark." We all are, after all, the hapless, if not helpless, subjects of the current moment in human history, which is largely defined by the logic of late capitalism. We, the authors, are no exception. At the same time, our personal experiences and empirical observations have shown us that there seems to be enough novelty in the recurring and expanding phenomenon that we study for it to deserve a neologism. We were ultimately relieved to see that our critical reviewer concurred with others that our concept does indeed have "theoretical legs." We would like to take this opportunity to thank all anonymous reviewers who took the time to read and comment on the manuscript, giving us very useful suggestions.

Other than empirical evidence and reviewers' assessment, we have had good fortune as participants of professional communities richly furnished with colleagues and venues that enable and reward our intellectual pursuits. We direct our acknowledgments toward recognizing, in the spirit of heteromation itself, the small, distributed, unsung, but crucial moments of invisible academic and non-academic interaction and activity that have woven themselves into this work. We also want to recognize the less proximate influences of the wider institutions that support research, whose nurturing and protection we often overlook. First and foremost, we express our gratitude to those who labor invisibly in the endless tasks of heteromation, for it was they who alerted us to the conundrum of heteromation before we had a word for it, and primed the long conversation that culminated in this book. We applaud the good cheer of these invisible workers, and their humility, humor, and eagerness to learn and share.

Much like all the little (virtual) rectangles under a curve that resolve into a particular shape, a life in the academy is composed of a stream of tasks that sometimes feel trivial and disconnected, but that do, in the end, take on a distinctive collective shape. In tribute to these tasks and the doers of the tasks, we thank all of the colleagues with whom we have published papers and books in the last several years, those with whom we have co-edited special issues of journals, those for whom we have written book chapters and who have written book chapters for us, those who write reviews for us or ask us to write reviews, those whose workshops we have attended, and those with whom we have organized workshops. You all know who you are. Your actions continually produce the vibrant community in which a book such as this can be written. The distinctive insights and knowledge of your research have found their way into the pages here.

We think this book can truly be described as interdisciplinary. It resulted in part from the eclecticism of the human–computer interaction community in which we both participate. We literally could not have written this book within our original academic homes because of their stricter genre conventions. We still identify with those disciplines, but are grateful to the human–computer interaction community for taking in wayward scholars whose intellectual interests do not quite fit into traditional disciplinary boundaries.

Our home institutions—Indiana University, Bloomington, and the University of California, Irvine—could not be better places to pursue the kind of research we do, and we honor their forbears, who understood that the world needs places where disciplinary perspectives can be transcended. As to our anonymous reviewers, you are the best! The nuanced critiques and grounded suggestions you offered with both tact and humor went a long way toward making the book considerably better than it would have been otherwise. The MIT Press, and, in particular, our editor, Katie Helke, as well as the co-editors of the Acting with Technology Series, Victor Kaptelinin and Kirsten Foot, provided careful guidance and encouragement throughout this project, for which we are very grateful. Chapter 8 was informed by joint work with our esteemed colleague Selma Šabanović, whose insightful understanding of social robotics allowed us to develop an especially interesting and important case of heteromation.

Various staff members in places that we have worked have made an indispensible, though invisible, contribution to our work. They provide invaluable professional expertise that makes it possible for us academics to focus on what we do. It is difficult to name them all, but we would like to thank them for their help and support.

Hamid's Acknowledgments

I would first like to thank my students in various courses, who patiently witnessed, and suffered through, the development of my ideas on the topic of this book throughout the years. I also thank the members of our research group CROMI for their creative engagement with these ideas, as well as those who have participated in our meetings in the last few years. More recently, I have benefited from conversations at the HCI/d group, the Marx reading group, and the "Working Subject" reading group in Bloomington, and in talks and conversations at IFK (Internationales Forschungszentrum Kulturwissenschaften) in Vienna. My colleagues at the Department of International Studies have heard me babble about heteromation here and there. I'd like to thank them all for their global perspective on these topics.

My very personal thanks go to Les Gasser, Harmeet Sawhney, and Jannis Kallinikos for numerous conversations on this and related topics throughout the years, to Douglas Hofstadter and Baofen Lin for their unparalleled care and kindness, to Michael Gasser and Mara Margolis for their passion for social issues, to Blaise Cronin and Betsy Stirrat for their creative companionship, to Reza and Florence Aslinia for their good hearts, to Deb Kremer for being there any time I need consultation, and to Stefano Raimondi and Christa Felder for their enduring friendship and their generosity in offering their beautiful home in Lechbruck am See, allowing me to work on the last few chapters of the book in the serene beauty of Bavaria. I'd also like to remember my dear friend Ali Halabian, a person with a deep concern for social justice, who passed away as this book was going to press.

Working with Bonnie throughout these years was a great pleasure, and in more than one way. Our paths crossed in an interesting moment in our intellectual careers, and the payoff came in how we could bring our different backgrounds to bear on issues that we both deeply care about.

My not-any-longer children Kaveh, Kia, and Taraneh, as well as their respective partners Leila, Pantea, and a yet-to-be-named lucky man, were major sources of inspiration and trepidation throughout this process. Observing them flourish but also struggle with the cultural, political, and economic challenges of the new world provided interesting "test cases" for my thinking. I'm sure they can discern their traces in different places in the book, and it is to them and their generation that I dedicate this book.

To Mahin, who shares my hopes and trepidations for our children, I don't have the right words to do justice to her grace, spunk, and selflessness, so I resort to Rumi for that:

> After my errands, to you I return;
>   Around your abode, I'll turn and turn!
> My vows to myself—are sealed in an urn;
>   Seeing your face again, they're fated to burn!
>
> —Translated by Maryam Dilmaghani

Bonnie's Acknowledgments

The PhD and Masters students I have been lucky enough to work with over the years have influenced the book in ways too numerous to mention, but which they themselves will recognize. Our TechDec lab meetings, and my qualitative methods and digital media classes, were stimulating sources of useful and interesting conversations on the work herein as well as related topics. The topics in chapter 12 grew out of work with my "Computing within LIMITS" colleagues, a hardy bunch that refuses the status quo and seeks concrete ways of addressing how things can be different. I am grateful for their perspectives and their courage. The Department of Informatics, and everyone in it, from office staff, to undergraduate and graduate students, to faculty, and department chair, have conjured a "powerful trouble" (in a good way) from the widely divergent skills, interests, expertises, and approaches of its remarkable members. I thank the undergraduates in my 2005 projects class who told me about social life in video games and sent me on an unexpected path that continues to feed my academic interests, as will be evident in this book, and is a great excuse for fun!

In celebration of invisible work with pervasive effects, I recall with admiration the cognitive psychologists and computer scientists who were open to the ways of anthropology and enabled my transition, earlier in my career, to the field of human–computer interaction. HCI is sometimes criticized as *too* open, but it makes an interesting and welcoming home for those of us who tend to color outside the lines. I thank these people from (formerly) foreign tribes who had the will to rework the tribal boundaries. Anthropology itself has been evolving, and there now exists a savvy group of young anthropologists industriously studying digital culture, and shaping how we understand it in exciting, provocative ways. It is wonderful to see them haul out old concepts like Lévi-Strauss's notion of bricolage, and deploy them to make sense of what is happening in the world of digital technology. In youthful, contemporary ways, they carry on vintage traditions that bring to bear anthropology's signature obsession with getting the details right so that the big picture can be right. I want to thank them too.

Hamid has been a redoubtable colleague on this journey with its many twists and turns. I have gained immeasurably from having had the great good fortune to work with him and to explore the issues in the book in ways I could not have done on my own.

Victor Kaptelinin deserves a special mention as a colleague whose sustained influence and support over many years have cast a clear light on my intellectual path. My co-editors at *Mind, Culture, and Activity*, with whom I am embarking on the project of putting into practice some of my beliefs about distributing wealth and knowledge (however small this drop in the ocean may be), have provided inspiration, as well as continuing interaction with the global activity theory community that has been my theoretical touchstone for many years.

My friend and neighbor Lynne Fox has patiently listened for quite some time to periodic reports of the progress of this book. An author herself, Lynne's kindness and empathy meant a lot. Russell Crispin generously offered frequent reality checks (and delicious home-baked goods).

I dedicate this book to my children. Anthony, Christopher, and Jeanette have been thoughtful, wise interlocutors on all phases of this project, and the best company in the world, as they always have been. The good vibes and lively conversation of their partners, Jaclyn, Mia, and Franco, always bring family gatherings to life.

The next generation—Cara, Lythia, Lila, and Lionel—are the ones who will bear the brunt, if that is the right word, of the long-term effects of what we speak of in this book. Though yet relative babes in the complicated woods of contemporary life, their already apparent curiosity and sparkle will see them through as well as anything can.

Prologue: An (Untold) Story of Economy and Computing

Modern technology is nothing short of magic. More magical, however, is human labor, ingenuity, and creativity. This book is about the relationship between these two magical phenomena—that is, the division of labor between humans and machines, and how it is configured and put to use by contemporary capitalism. The socioeconomic changes of capitalism in the last two hundred years find close parallels in the growth of technology. Far from being a coincidence, these parallels demonstrate the interlocking development of economy and technology.

The computerization of the economy in recent years has transformed relations of human persons and digital machinery, systematically moving a large majority of people toward essential but marginal roles. Much of the work undertaken in the new division of labor is hidden, poorly compensated, or uncompensated, and naturalized as part of what it means to be a "user" of digital technology. This work is not recognized as the economically valuable labor it actually is. Here, capitalism does what it has always done, industriously discovering new mechanisms for accumulating wealth, and altering the work and lives of those participating in its growth, often in asymmetric and unfair ways. In this book, we examine instances of this kind of participation in economic production, which we dub "heteromation" (Ekbia and Nardi 2014). Heteromation is the extraction of economic value from low-cost or free labor in computer-mediated networks. *It is, in essence, a new logic of capital accumulation.* The wide variety of heteromated labor discussed here shows how the masses are taken up into networks, the value they generate when they get there, and why this value remains invisible.

P.1 Background

Marx and Engels pinpointed labor, or practical activity, as the very source of anthropogenesis, the momentous shift from primate to human society. Human labor, for them, is also the source of all value—the ultimate mediation. In the last three hundred years, economists, and to a lesser degree, sociologists and anthropologists, have taken great pains to theorize the role of labor in economies. In writing this book, however, we came to an interest in labor not through theory, but in our own encounters with digital technology. In 2005, Bonnie had purchased the game World of Warcraft for her studies of online social life. Upon opening the box of disks that contained the game software, she was puzzled to see that what appeared to be a complicated game arrived with no documentation. She soon discovered that World of Warcraft was even more complicated than she thought, but that learning was amply supported through in-game social interactions, YouTube videos, user-produced game guides, forums, blogs, podcasts, user-produced software modifications, and even user-generated statistical analyses of the algorithms underlying game mechanics that allowed players to understand, with considerable rigor, how to play the game more effectively. This puzzlement represented a bigger phenomenon that begged explanation.

That phenomenon is also prevalent in AI, where a long tradition of hyped discourse has attributed extraordinary feats to computer artifacts, portraying them as "smart," "affective," "social," "ethical." The socioeconomic and philosophical drivers of this tradition have long been an interest of Hamid's, who saw the discourse as a means of dealing with an inherent tension in AI between its engineering ethos and scientific aspirations—a tension that, in turn, puts into high relief deep-rooted dilemmas of modernism (Ekbia 2008). If, in the discourse of mainstream AI, chess programs are described as champions of human thinking, meager algorithms as capable of understanding human language, and simple robots as displaying affect and emotions, this has less to do with technological magic than with the human ingenuity that has gone into the design of these systems. And it has more to do with the insatiable appetite of mainstream media and Hollywood for sensational stories than with genuine scientific understanding, and everything to do with massive military funding that seeks robotic replacements for human warriors. It is the tension between hype and hope—between the drivers of this discourse and the enormous capacities of computer technologies for improving human wellbeing—that poses another fascinating puzzle.

In discussing these puzzlements with each other, we began a conversation about the roles of humans and machines in digital activity that then led us to return to Michel Foucault, Karl Marx, and other theorists of modernity to ask how innovations in computing affect labor, and, more broadly, life. In particular, we asked three questions:

1. What are the social and technological processes through which economic value is extracted from digitally mediated labor?
2. What is the nature of the value created in this process to allow capital to continue its necessary expansion?
3. How are people incited to participate in this process?

We gloss these, respectively, the process, the outcome, and the stimulus questions. We address these questions in this book.

We conceive our audience to be readers who, much like ourselves, are tech savvy and alert to the nuances of social life, but less aware of the dynamics of capitalism. This book, has, as such, involved a set of "detours" through which we extended our usual concerns to explore new terrain that seemed increasingly relevant. To take these detours, we drew on traditions and literatures in the social sciences, technology studies, economics, informatics, and beyond. Adopting Foucault's account of modernity as the purveyor of various disciplinary mechanisms, we have, in the past, examined the shifting relation between humans and machines as one that inserts human beings as instruments into sparsely structured technical systems (e.g., computer games, social media) that operate on a totalized logic of engagement (Ekbia and Nardi 2012, 2014). We agree with philosopher Brian Holmes that a transformation is underway from Foucault's disciplines to a new order in which we participate willingly: "Rather than coercive discipline, [there] is a new form of internalized vocation, a 'calling' to creative self-fulfillment in and through each … project, that will now shape and direct … behavior" (Holmes 2002). It is this new order we seek to problematize, in Freire's sense of "generating critical consciousness … to perceive the social, political, and economic forces that influence human lives" (Freire 2002). We do this in order to understand more deeply how digital technologies penetrate our lives, with the intention of cultivating a critical consciousness about technologies' increasing influences and impingements.

P.2 Looking Back: A Parallel History of Computing and Capitalism

Our exploration begins with a historical overview of the co-development of technology and economy by focusing on the division of labor among human beings, and between humans and machines. We contend that the contemporary capitalist economy is significantly defined by the way it mediates and captures human activity in networks. Modern societies have constructed a remarkable infrastructure that interposes itself in daily life in a layered fashion of continual engagement: through physical actions (clicks, swipes, key presses), identity representations (logins, profiles, user names, preferences), digital practices (checking email, posting to Twitter, reading Reddit ...), and life projects (work, leisure, family, friendship). These actions, representations, practices, and projects generate economic value via the scaling possible in massive networks. By assembling and maintaining these networks, capital taps into reserves of human labor that can be strategically leveraged. The majority of us willingly participate in a set of amalgamated digital activities that yield economic value to others. A darker, more coercive side of computing hovers out of sight too; coercive mechanisms produce unwanted outcomes that exploit our exposure in networks or intrude into our time and resources. Issues of "privacy" nod to some of the concerns, but fall short of grasping the underlying processes that enable and generate them. We examine these processes along with processes of consent and participation that shape digital activity.

P.2.1 Predicaments and Possibilities

In explaining the mechanisms that induce participation, we probe "social predicaments," locating them not as individual psychological problems (e.g., internet addiction), or purely social phenomena to be studied at the level of culture (e.g., as affective investments), but as complex developments issuing from the intersection of economy, technology, and social life that materially affect individuals. Predicaments have shifted, along with the changing conditions of capitalism. They are not to be understood as neat problems with neat solutions, but broad tensions, conflicts, disturbances, and existential realities that cannot yield to precise "fixes." Predicaments become part of the everyday life world, and we live with their ambiguities and uncertainties.

It would be imprudent of us, and of anyone for that matter, to suggest that predicaments can be easily resolved. That is not, therefore, what we seek to do here. What we want to accomplish instead is to provide a clear understanding of the predicaments and how they are generated through our collective and individual actions—collective because the choices that we as communities and societies make give rise to these predicaments, and individual because every one of us enacts the predicaments through our actions and behaviors. While it is true that social change emerges from the decisions and actions of various actors along the way, and that no particular group can be held solely responsible, it is also true that the direction of change was not inevitable in the past, nor is it inevitable in the future. The historical truth of this innocuous statement might be obscured by the accelerating speed of change in the last few decades. Younger generations have not experienced life without laptops, smartphones, and music players, nor have they experienced accessible and affordable education, secure jobs, and stable lifestyles. Older generations, on the other hand, playing catch-up with the new technologies, do not quite appreciate the sense of attachment, engagement, and participation associated with these technologies in the minds and lives of the young. Neither group may fully grasp the incurred costs or the potential affordances of engagement and participation, nor may they understand how the two arenas, technology and economy, relate to and feed off each other.

People's own aspirations for active engagement with technologies inside and outside work have provided ample opportunity for the invention and introduction of an array of new products and services. On close scrutiny, many of the innovations in digital technologies that have propelled consumption—e.g., mobile technologies, social networking, gaming—can be understood as ways to address predicaments that have intensified during the last few decades. In this context, predicaments may be thought of as fundamental existential conditions of contemporary capitalism. We do not continually experience them as acute struggles, but we are embedded in complex conditions from which we cannot easily extricate ourselves.

Contemporary capitalism itself rests on a central predicament—the need to continually generate growth at about 3 percent a year (Harvey 2010). This is a colossal dare, and capitalism has proven endlessly inventive and adventurous in meeting its defining challenge. Economic growth has followed from the sorcery of capitalism: a 16-fold rise in the per capita income from $3 a day in 1800 to $100 in the early 2000s for most of Europe and North America (McCloskey 2009). This level of accomplishment, according to the proponents of capitalism, is what makes it uniquely different from all socioeconomic systems of the past. And, in a way, this is right: people in much of Europe, North America, and select countries of Asia and the Pacific, live richer, healthier, and longer lives, on average, than their ancestors a few generations ago, and they are more informed, educated, and entertained. Economists disagree in explaining the drivers of these changes, suggesting expanded trade, diversified resources, industrial thrift, facilitated transportation, incentives and institutions, and a host of other parameters. The true story is probably more complex and multicausal than any of these, but the views converge on what can be broadly described as *innovation*, which according to economic historian Deirdre McCloskey, "has elevated people, in more than goods alone" (2009, p. 2).

Capitalist innovation has created great opportunity for individual prosperity and systemic resilience. The most recent manifestation of these is the range of opportunities provided by computing technologies for economic productivity, cultural self-expression, political participation, sharing of ideas and projects, educational opportunities, even the development of spirituality and ethics. In this book, we are as interested in the opportunities for creativity and participation through digital mediation that people perceive and seize, sometimes with astonishing verve and ingenuity, as we are in the predicaments. Just as technology affords ways to address social predicaments, it also enhances opportunities and possibilities to harness the positive forces of modern life.

It is sometimes difficult to logically differentiate predicaments and opportunities—they mutually produce and reinforce one another—but it is important to recognize, as a separate impetus, the initiative and resourcefulness people summon to create and pursue an array of diverse opportunities. The interplay between the pressures of predicaments and the liberatory potentials of digitally mediated opportunities should be examined and untangled for modern individuals to be able to makes sense of their condition. Uptake of opportunities is always entwined with more reactive strategies of calculative control responsive to predicaments. Sometimes the two happily merge (as work and play can merge when we enjoy paid labor), and this possibility contributes to the committed nature of contemporary digital experience.

P.2.2 Technology as Response

Opportunities tend to develop in the midst of personal or social plans and projects imbued with the particularities of personalities, individual histories, local circumstances, and unpredictable confluences of events and conditions. To be a person—a worker, student, parent, citizen, friend—one cannot escape the predicaments, but one can run with opportunities that spark or even compel activity. As we will discuss in chapter 1, the history of digital technologies itself loosely mirrors the duality between predicaments and opportunities. Vannevar Bush (1945) suggested that with digital technology, "Man's spirit should be elevated." Throughout history, technology has propelled utilitarian programs of economic production and military conquest, being little concerned with elevating anyone's spirit. Bush's comment refigures constructions of computers as, for example, aids to improvements in manufacturing operations or efficient tools for organizing data, suggesting higher purposes. Digitally mediated opportunities, with their infusions of personality, spirit, and history, seem likely sources of ferment and development, with the potential to destabilize the inevitability of particular changes we are so often assured we cannot escape. Anyone reading this book has used technology to their advantage, probably for many years, and with considerable success.

Our argument is, therefore, quite simple: digital technologies are responses to the predicaments and opportunities of capitalism. We locate technologies as processes within a dialectic involving predicaments and opportunities, on the one hand, and the growth and development of specific digital technologies, on the other. In a system as complex and indeterminate as capitalism, there are no simple causal paths to be traced, but we should not decline to notice the unmistakable propensity of capitalism to locate profit-making possibilities wherever they lie, and seize them aggressively. If we fail to follow the eventual trajectories of those seized possibilities, we fail to see how the system in which we are so thoroughly enmeshed operates, and how it affects us.

While others debate whether digital technologies are mere *tools* or *drivers* of social change, we regard them largely as a response to the dynamics of socioeconomic change. Technologies cannot constitute exogenous forces entering the economy from unspecified sources—an unspoken assumption of the tools/drivers perspectives. The tools narrative suggests that technologies simply appear, to be picked up if they look good for a particular purpose (see Hornborg's critique, 2001). The driver story correctly points to the shaping power of technology, but neglects to consider the agency of individual persons responsive to predicaments and opportunities. We will argue that the capitalist economy has created the conditions for the possibility of computer technologies as they have developed so far, and, at the same time, computer technologies have provided part of the solution to problems of capitalism.

P.3   Theoretical Orientation: The Political Economy of Computing

We originally began this book under the influence of Michel Foucault, especially our reading of *Discipline and Punish*. It seemed we could write a book about digital control, picking up where Foucault left off, bringing his insights forward to the contemporary moment. Once immersed in our materials, however, the differences between his world and ours precluded easy passage from prisons and schools to the blooming, buzzing confusion that is life lived digitally. It became apparent that icons such as the panopticon, with its graceless intrusions, would tell us little we would need to know about digital control's light, elegant, delicate, implicit, hidden, continuous, dispersed nature. The astonishing array of nearly invisible digital mechanisms—metrics of monitored work environments, peer pressure in social media, governmental and corporate massing of personal data, increasing medical surveillance afforded by wearable technologies, the filter bubble narrowing our field of view and trivializing us through overexposure to kittens and goofy videos—would need fresh treatment. In particular, the committed engagement of contemporary subjects with digital technologies upsets the apple cart of the coercions of Foucault's imposed disciplines. These disciplines, with their us-them clarity of subjects and masters, did not compute when applied to the intricacies of the current digitally infused context. Digital control and participation are much more of a dance, and a tangle.

The perspective presented in this book can be broadly understood as the "political economy of computing" (Ekbia and Nardi 2015, 2016). This perspective takes into account socioeconomic drivers of change in the relationship between humans and machines, much as Charles Babbage, right from the beginning of the age of computing, sought to do in his book *On Machinery and Manufactures* (1832; see chapter 1). Rather than focusing on the characteristics of the technology, on individual dealings with machines, or on organizational implications of technology, this perspective views technological change within the frame of historical developments, socioeconomic systems, the division of labor, legal and regulatory frameworks, and large-scale government policies and agendas. Working from the vantage point of historical forces and conditions allows us to consider issues of social welfare, employment, economic rewards and incentives, power and politics, governance structures, safety and security, and the technologically mediated relationship between life, labor, and leisure.

A word on language is in order here. Capitalism, which is a central notion in this book, is not a fixed system that acts on individuals and the society from outside, determining their choices and actions in a causal manner; rather, it is shaped and transformed by the personal and collective actions of those same individuals. This reality, of course, does not mean that individuals are "free" to make choices on a whim, as neoliberal thinkers suggest. The relationship between the two is one of co-constitution. To capture this two-way dialectic influence, throughout the book we use a symmetrical language of the sort "Individuals do X" and "Capitalism does Y," knowing full well that neither one has absolute agency and control over their choices and actions.

In the same vein, we sometimes refer to the "masses" taking action, responding to situations, or being subject to certain types of predicaments. In doing so, we do not mean to imply a monolith that acts in unison. Nor do we mean to erase class, gender, ethnic, and cultural differences. We strongly believe in the meaning and material significance of these distinctions. For our purposes here, however, the most relevant distinction is between those who produce value for others through their labor, broadly construed, and those who extract value from the labor of others. We refer to these two groups as, respectively, the "masses" and "capitalists," acknowledging at the same time that these are not black-and-white and stable categories; rather, they slide and shift both in individual and historical senses (see chapter 2).

Lastly, the concept of "network" is central to our account, as it is to many other accounts of contemporary societies. Because of its currency, the concept is used with very different semantic connotations in various discourses. On the one hand, there is the narrow discourse of "networks" as a recent phenomenon enabled by digital and electronic communication. On the other, there is the broad meaning of the term as a kind of social structure that has superseded other historical structures such as class. The narrow meaning pervades technical literature on social media and social "networking," while the broad meaning can be found in a certain brand of academic work that sees everything through the lens of "network analysis" (e.g., Barabási 2002), or in grand narratives of the "network society" that gloss over issues of class and class conflict (e.g., Castells 1989; see Schiller 2014).

We seek to avoid these tendencies, recognizing the historical reality of networks as one possible form of social structure that precedes computer technologies but that has nonetheless grown in prominence because of the widespread adoption of these technologies, without replacing alternative forms such as social classes and institutions. Our use of the concept of "network" is, therefore, akin to that of Boltanski and Thévenot (2006) and Boltanski and Chiapello (2005) as one among various polities in contemporary societies. In particular, we speak most often of people acting within computer-mediated networks, rather than networks as technical objects, ego-centered personal social networks, or generic superstructures.

P.4 Theoretical Landscape

Masses of people participate in computer-mediated economic activity in global networks. Embeddedness in networks is usually typified as either (1) celebratory visions of "crowds" coming together in social media revolutions, peer productions, free and open software collectives, projects of citizen science, gamified pedagogies, and so on; or (2) dystopian visions of masses serving as instruments of capital, often ensnared in obsessive, even "addictive" relations to technology, forming a standing army of oppressed precariat labor. In our view, neither imaginary is completely wrong, but each misrepresents and undertheorizes—one unjustly rationalizing the extraction of economic value as fair play, the other declaring universal exploitation! Neither deeply analyzes the processes and mechanisms by which the participation of the masses occurs, or the subtleties and ambiguities such participation entails. We seek to uncover these processes and mechanisms.

In developing our ideas, in addition to classics such as Karl Marx, Michel Foucault, and Max Weber, we have been inspired by the work of Luc Boltanski and Ève Chiapello (2005), Richard Sennett (2007), David Harvey (2010), George Caffentzis (2013), Luis Suarez-Villa (2012, 2015), Dan Schiller (1999, 2014), and others. Approaching digital technology more broadly than in our previous work, we engage the economy as an analytic object to discern the processes and trajectories of the digital technologies we have studied for the last couple decades. Foucault's influence is apparent in our attempt to identify the drivers, mechanisms, and instruments of discipline and control in contemporary societies. Although what we explicate departs in character from what Foucault described in his work, the nature of our inquiry remains indebted to him and his methods.

Despite the influence of these literatures on our thinking, the bodies of work we have discussed so far are largely silent on issues of technology, leaving significant questions to be explored. A similar statement can be made about Science and Technology Studies (STS), which, one would hope, should provide another reference literature for the kinds of questions that we deal with here. With a few exceptions (e.g., the work of Harry Collins on expert systems and Lucy Suchman on situated action), and for reasons that are beyond our concern here, the STS community remained disinterested in issues that have to do with computing for a long time. A cursory overview of the major venues of STS in earlier years would demonstrate much more focus on the life and environmental sciences (biotechnology, genetics, health) than computing. Being critically aware of this (Pinch 2010), the community has appeared to pay more attention to computer technologies recently, as evident in the programs of its main venues, such as the 4S Conference. To the extent that STS has engaged with computing technology, however, it has not shown much interest in the political economy of computing, and related questions of labor, value, and capital accumulation. The emphasis in actor-network theory on the agential *symmetry* between humans and machines, for instance, might have in fact blurred a more basic *asymmetry*—namely, crucial differences between them in regard to the question of labor, diluting the fact that only humans can produce economic value (see Star 1990 for a similar observation on the question of power in actor-network theory). Labor, in our view, is a uniquely human capacity, which accounts for the production of economic value in capitalism. Our own use of the term "division of labor between humans and machines" should be understood as metaphorical shorthand, for that matter.

Questions regarding labor and computing are taken up in recent literatures concerned with the production of economic value in social media, creative media, and microwork systems such as Amazon Mechanical Turk and CrowdFlower (Terranova 2000, 2003; van Dijk 2009, 2013; Ritzer and Jurgenson 2010; Fuchs 2011; Arvidsson and Colleoni 2012; Scholz 2013; Comor 2015; Robinson 2015; Maxwell 2016). We draw from this important literature, but go beyond it—heteromation succeeds as a logic of accumulation because it leverages so many different types of labor, deploying varied mechanisms (beyond those of social media and microwork) to stimulate participation and extract economic value. We analyze microwork and media, but we have in our sights a capitalist dynamic that pervades activity in global mediated networks—this dynamic is not the outcome of a few innovative platforms. We thus examine heteromation in activities of software development, design, social robotics, self-service, citizen science, banking, search, and others. Our objective is to emphasize the continually spreading nature of heteromation as it penetrates myriad forms of human activity across networks.

This objective brings us to another category of literature shaped around the sociocultural, psychological, and ethical implications of computing. Situated within information and communication studies, this literature encompasses, among other things, writings on ubiquitous computing, social computing, computer-mediated communication, computer-supported collaborative work, human–computer interaction , and social informatics. While valuable contributions in their own right, in our view, these works, with a few exceptions, commonly fall into one of three categories. The first category is celebratory accounts that depict computer-enabled developments of the last few decades as empowering, participatory, and productive for the masses. With artificial intelligence, the internet, and social media as focal topics, these portrayals promise an even playing field that can nurture individual talents and aspirations, enrich communities, produce machines with feelings and consciousness (Wozniak 2011), and even grant humans eternal life (Kurzweil 1999).

A second category is dystopian and alarmist narratives that construct the dull image of an inevitable future where digital technologies will generate a machine-addicted, coerced, and conformist society conjured in the minds of an elite group of smart, rich, and powerful individuals and institutions (McChesney 2013; satirically, Kirman et al. 2013). Sometimes, based on a critical understanding of the economic realities of technocapitalism (Suarez-Villa 2012), these accounts tend to emphasize the growing influence of corporations, particularly major banks and financial institutions, over the social and political life of contemporary societies. Others (e.g., Fuchs and Sevignani 2013), coming from a Marxist tradition, discuss "digital labor," examining the changing character of value production in capitalist economy, the role of "users" in the process, and the distribution of rewards. A broad debate is shaping up around these topics, with commentators of diverse persuasions taking part (Proffitt, Ekbia, and McDowell 2015).

A third common position of "technology apologists" is discernible among those who tend to provide detailed, nuanced narratives but who are either indifferent to social and economic injustice partly incurred as a result of ubiquitous use of computing technology, or who seem to understand their role as accepting and explaining away the status quo or endorsing it through proposed micro-adjustments in technology and/or organizational practice. To our chagrin, common examples of the this group can be found throughout academia, which was once the bastion of critical thinking that provided viable alternatives to both ends of the spectrum. Apologist narratives occur, with weak or strong streaks, in much of the literature on ubiquitous computing, human–computer interaction, computer-supported collaborative work, and elsewhere. Significant exceptions include arts-inflected perspectives; for example, "speculative design," an approach that deliberately separates itself from the marketplace in order to "explore alternatives to our current model of capitalism" (Dunne and Raby 2013, p. 12; see also Penny 2016), as well as work on sustainability and human–computer interaction that challenges capitalist practices such as planned obsolescence (Blevis 2007; Remy and Huang 2014).

We seek to steer clear of the apologist position, and offer a balanced and realistic image that identifies the celebratory and dystopian scenarios as extreme points on a broad spectrum of options and possibilities. While recognizing the slight likelihood of the realization of either vision, we believe that a more nuanced account of contemporary socioeconomic displacements based on social realism (Kling 1996) has a higher potential for enlightenment and empowerment. There is a great deal of complexity and confusion surrounding technology, economy, and culture, with most of us being inundated by a wave of change that feels like it is beyond our grasp and control. Critical analysis should aim to inject some sense into this confusion.

We are well aware of the challenge of articulating a balanced perspective. This is *our* predicament, which reflects the bigger predicaments of contemporary life to which we alluded, and provides the ambivalent situation in which everyone, including critics, is embedded (Wagner 1994). Rather than ignoring or minimizing the challenge, though, we want to face it head on because we find this approach as the best guarantee to shun the positions of a return to an idealized past, or unbridled enthusiasm for technologically driven social and economic change, or dark visions of entrapment in the Matrix, or even the end of civilization brought on by technoeconomic apocalypse.

P.5 Looking Ahead: Critique, Nostalgia, and Ideology

Capitalism depends for its vitality on the participation of the masses, as laborers in the production process, and as consumers in the market. Unlike economic systems of the past, such as slavery or feudalism, the masses are neither possessed as chattel, nor can they be forcibly coerced into participation. As such, they have the capacity to resist, and this is the positive source of the appeal and power of capitalism compared to other systems. On the flip side, this capacity for resistance comes as a challenge to capitalists who need to incite the masses to participate either through control mechanisms, consent-producing practices, or a combination of the two (Caffentzis 2013). Digital technology has become a critical means by which control and consent are produced and managed. A realistic approach should consider both of these aspects. Part of the goal of this book is to promote reflection, to trouble uncritical acceptance and evaluation of technological development by widening the frame in which we employ and assess digital technology. We need a space of reflective activity where we can become aware of anxieties and enigmas, temporarily escaping the normalcy of the everyday.

Gilles Deleuze (1992), comparing earlier disciplinary societies and more contemporary ones, said: "There is no need to ask which is the toughest regime, for it's within each of them that liberating and enslaving forces confront one another." His observation leads to a key point that constitutes a central motif of this book concerning the purpose of critical analysis. The critique of digitalization provided here should not be understood as a call to go back in history to a time when processes of control and oppression were more visible, straightforward, and hence graspable. Complex and intricate mechanisms of change call for appropriately complex tools of analysis and understanding. We take this to be the challenge for us and for any critique of the status quo. We do not operate on a nostalgic view of a bygone past, where things allegedly worked more simply, more beautifully, and more humanely. We are of the belief that such nostalgic sentiments, to the extent that they become operationalized, feed into conservative, if not reactionary, attitudes and policies.

By the same token, we do not believe that every new technology makes us smarter, faster, better, more modern, more progressive. Nor do we expect that humanity is advancing along an irreversible path toward Machine Singularity, a vision lodged in our (un)consciousness as far back as the nineteenth century, when it was suggested that machines "may … think of a plan to remedy all their own defects and then grind out ideas beyond the ken of the mortal mind!" (Thornton 1847). Technologies rarely offer genuine fixes to complex human problems (much less remedying their own defects), and they invariably spring open Pandora's boxes of externalities and unintended consequences. The internal combustion engine, to take but one example from among many, has ended in devastating ecological and social consequences that have proven stubbornly resistant to repair and remediation. We thus aim to avoid both nostalgist and dystopian perspectives, and to instead provide an account that reveals the mechanisms, processes, and dynamics of actions within the current capitalist system, with projection forward to the future.

In that spirit, our allusion to Karl Marx and his theory of capitalism throughout the book should not be understood as pushing an ideological agenda. In our minds, Marx, as a sociologist and philosopher, was one of the first thinkers who had the analytic insight to identify the subtle mechanisms of value generation in the capitalist economy. While capitalism has changed drastically since Marx's time, many of the insights and analytic tools that he developed still apply and remain relevant today (Ekbia and Nardi 2015; see also chapters 2 and 3 of this book). One can describe these as Marxian analysis to differentiate them from Marxism, which relates to Marx's theory of historical materialism. Marx was also one of few social theorists who gave a special place to machines and technology in his thinking. His law of the tendency of rate of profit to fall (TRPF), which described the seeming paradox of technologically driven rising productivity and decreasing profit (Marx 1990), still rings true today. Heteromation can actually be understood as a capitalist innovation to deal with the tension that derives from this paradox.

Being non-ideological, however, does not mean that we should shy away from incorporating politics in the conceptual analysis of technology. Politicizing, unfortunately, has been rendered a taboo in our society by those who are most political in their intent. The pressure to be non-political in one's views and theories is just another way of silencing dissenting voices, in the same manner that any mention of class difference is frowned upon as "class warfare" by those who are most combatant in protecting their own class interests. Technologies are inherently political (Winner 1986), as are design choices, and pretending otherwise is not going to erase politics from technological arrangements.

To speak of technologies as inherently political should not be taken as the expression of a conspiratorial view. Far from this, we do not subscribe to those perspectives that consider capitalism and its developments as the (bastard) brainchild of a group of greedy capitalists who are in the business of constantly plotting and devising new mechanisms of exploitation. Nor do we believe that capitalists, even if they had the desire, are capable of that—that would give too much credit to capitalists. Rather, as students of Marx and Foucault, we believe that change in capitalist societies (of the kind we examine here) is driven by the inherent dynamics of the socioeconomic system, which include, among other things, mechanisms of value extraction, market coordination, consent building, participation, and governance.

To support the arguments, our narrative draws on publicly available reports and data, personal narratives, as well as our own studies and observations of populations of video gamers, microworkers, YouTube creators, and other social groups. The book is not based on a single comprehensive empirical study, nor does it hew to any single theory, but attempts to provide an interpretive account with vivid and grounded images of the kinds of experiences and predicaments people face in the heavily computerized environments of today. Indeed, given the rapid rate of change, and the unprecedented complexities and affordances of digital technology, we are convinced that broad interpretive treatment is the way forward at this time. We must, in a pre-theoretical spirit that draws from several theories, but aspires primarily to put some complicated ducks in a row, prepare ourselves for theory. Just as Darwin painstakingly performed an essential cataloguing of minute observations as a necessary step toward his broader goals, we aim to shine a light on patterns of technology and economy that we believe are only glimpsed here and there, in scattered literatures. The concept of heteromation allows us to do this to a great extent. We hope that the reader can identify with, and relate to, the pertinent cases and examples through their own personal experiences and those around them. Should this happen, it would be a step toward social awareness and social action, which we consider to be the function of critique.

P.6 Outline of the Book

The focus of the book is on heteromation as a new logic of capitalist accumulation. Each type of heteromation involves specific human capacities brought to bear in the extraction of economic value. Each type of heteromation also requires and engages specific mechanisms for securing value and making it invisible at the same time. In part I, we consider the emergence and development of these mechanisms within the historical displacements of capitalist societies in the last century or so. Chapter 1 provides a short history of computing pertinent to issues of heteromation, with a focus on the division of labor between humans and machines.

Chapter 2 examines logics of wealth accumulation, as articulated by some of the thinkers and commentators of the last two centuries. Chapter 3 describes societal displacements, with a focus on value generation, which remain at the core of capitalism. We illustrate the dynamics of change that have allowed capitalism to continuously reinvent itself to guarantee the securing of economic value, and the innovative technology that has been a key part of generating value. This chapter discusses our increasing personal engagement with technology beginning in the post–World War II era. Chapter 4 continues this discussion, examining stimulus mechanisms, i.e., what incites the masses to participate in computer-mediated activity. In the absence of direct coercion, these mechanisms may be said to consist of possibilities and predicaments. Possibilities are positive drivers that *pull* people toward certain types of behavior through a combination of financial, psychological, and cultural forms. Predicaments, on the other hand, are negative drivers, such as precarity, separation, futility, and monotony, that *push* people toward those same behaviors. We see a direct relationship between these and the participation in computer-mediated networks by individuals from virtually all socioeconomic backgrounds, even homeless people (see, e.g., Woelfer and Hendry 2011).

In part II, chapters are organized according to different kinds of heteromated labor based on the human capacities they engage. These capacities naturally co-occur in complex human activity, but each chapter chooses a single capacity as its focus. Our use of the categories is suggestive, intended to point to the range of human abilities heteromation leverages, and how flexibly it manages the extraction of value.

Chapter 5 discusses *communicative labor*, covering the wide spectrum of contributions, including the user-generated content of social media, most vividly captured by platforms such as Facebook and Twitter, where value is extracted through the communicative practices of daily life. We show how the network mechanisms of digitally mediated environments hide the contributions of masses who find themselves separated and in need of connection.

Chapter 6 describes *cognitive labor*, including activities such as microwork and self-service. Cognitive labor rewards participants with meager monetary compensation, or in the form of small but useful functional outputs (such as receiving cash from an ATM), which tends to hide the economic value of such activities.

Chapter 7 concerns *creative labor,* which is similar to cognitive labor, except for its strong intellectual component, as seen in gaming environments, literary productions, graphic design contests, YouTube content creation, and so on. What hides the economic contribution here is the strong sense of personal worth that derives from human creative capacity. In this sense, creative labor often (though not always) concerns the possibilities inherent in computer-mediated networks rather than the predicaments.

Chapter 8 considers the *emotional labor* that engages human beings as social agents. Social robots, for example, although ostensibly paragons of automation, actually require significant human mediation. For example, elderly individuals who interact with these robots show interest, but they also need a great deal of human hand-holding to guide them to interact with the "social" robots. In Brazil, correspondent banking for residents of remote rural areas requires socially adept humans to help low-income people access banking services (Bailey, Diniz, and Scholler 2014; Bailey et al. 2016). The economic value extracted from such labor is hidden within paid jobs; workers do not receive extra compensation, although they have added new skills and tasks to their repertoires. This labor engages the highly developed, caring person produced through a lifetime of social experience.

Chapter 9 examines *organizing labor,* in which people organize themselves into effective collaborative groups. Citizen scientists, for example, are often called upon to organize or participate in events and experiments to collect samples, raise awareness, or simply clean up a habitat—tasks that are too expensive or impossible for machines to perform. The value of this labor is hidden by participants' keen desire for positive social experience. These varieties of heteromated labor are shifting the socioeconomic, political, and cultural landscape of life in visible and invisible ways.

In part III, we examine some of these shifts, with an eye on the future. Chapter 10 considers the mechanisms that incite participation in heteromated labor. Chapter 11 investigates shifts in styles and modes of governance. The growing prominence of heteromated types of labor, for instance, affects our notions of welfare, employment, retirement, insurance, training, and education.

In chapter 12, we sketch a utopian vision that imagine the technological mechanisms deployed in heteromation refigured on a new economic footing that would bring end users more fully into the prosperity of capitalism. Heteromation, after all, is not determined by engineering or computing principles, but by modes of production in the contemporary economy. Modes of production can shift dramatically, as they do at consequential moments in human history. And perhaps they must shift, given global instabilities, which could open favorable new possibilities for work and life.

The epilogue pulls these different threads together in an attempt to provide an overview of the intellectual trajectory of the book, including the "detours" that needed to be taken in order to address our underlying question about relations between people and machines. Appearances notwithstanding, we will see that answers to this question are anything but obvious.

I
Looking Back

1
Heteromation: A Revised (His)story of Computing

1.1 Introduction

Capitalism is, generally speaking, the pursuit of unlimited accumulation of capital, expressed in Benjamin Franklin's famous dictum, "Money can beget money, and its offspring can beget more, and so on." The notion of capital has been reified in most people's minds as money or as a tangible asset. David Harvey (2010, p. 40) reminds us, however, that, "Capital is not a thing but a process in which money is perpetually in search of more money." The relentless demand to accumulate wealth impels constant, energetic economic activity (Boltanski and Chiapello 2005; Sennett 2007; Wright 2009; Suarez-Villa 2012, 2015), giving capitalism its uniquely restless, driven character.

Some of the most highly capitalized and powerful enterprises in the world today are technology companies—the owners of indispensable infrastructures within which systems of commerce, communication, transportation, science, education, healthcare, entertainment, government, and social life are embedded (Morozov 2011). The relationship between computing and capitalism, however, has deeper historical roots as far back as 1801, when Joseph Jacquard demonstrated that a loom controlled by a series of punched cards with coded designs could create beautiful, complex fabrics such as the damasks and brocades traditionally woven on hand-operated looms. Changing designs was as simple as changing cards, and demand for the fabrics portended significant profits by offering more designs at lower cost. Inspired by Jacquard, Charles Babbage wrote about computing machines and their essential role in economic development in *On the Economy of Machinery and Manufactures* (1832). Although better known for the design of the Analytical Engine,

Babbage's keen understanding of the economic potential of computing was a critical contribution to explaining a rapidly changing economy, and an important influence on Marx and other economic theorists (Caffentzis 2013).

The Jacquard loom engaged the principle of automation to increase profits. There is a limit, however, to the level of profit derived from automation, stemming from technical reasons having to do with the shortcomings of machines compared to humans. However, there is also a more important economic reason: human labor is the key source of value creation in capitalism, as Marx insightfully showed in his theory of surplus value. Total elimination of human labor, therefore, is not a viable option, although it might be a capitalist dream to fully automate economic processes. To automate or not to automate—this is a central question of capitalist economy. It is largely in dealing with this question that computing technology has evolved within capitalism. Along with this evolution, three distinct but intercalated phases of human–machine relations have emerged: automation, augmentation, and heteromation.

Automation intensified following World War II, concurrently with the beginning of the widespread deployment of digital computers. Phenomenal new machines enabled corporations and bureaucracies to support the most complex organizational operations that business and government had ever had. Augmentation, by contrast, drew on a more personal, subversive, individualistic element in the culture of the 1970s, and propelled the transition to the diffusion of personal computing into everyday life, creating a huge market of millions of individual consumers (Campbell-Kelly and Aspray 1996). Computers augmented the capacities of ordinary people, who, at this time, became "computer users." These users could now find information, communicate with one another, crunch numbers, draw pictures, play computer games, and even learn to program the computer to do whatever they might wish. Lastly, heteromation moved the empowered person of the augmentation phase to a more marginal role, functioning at lower capacity on the margins of the machine or the organization (Ekbia and Nardi 2012, 2014; Ekbia, Nardi, and Šabanović 2015).

Heteromation has myriad forms, and we will discuss the appropriation of communicative, cognitive, creative, emotional, and organizing labor of human beings in computerized environments. Heteromation extracts economic value from *uncompensated* or *low-wage labor*, inciting participation through an intricate set of mechanisms comprised of social and emotional rewards, monetary compensation, and coercion. Generating this value doesn't cost capital much, yet it summons intelligent human labor from the masses across global networks of billions of nodes. "All the work, without the workers," as a character in Alex Rivera's 2008 film *Sleep Dealer* says, as he labors at the job of controlling American construction robots from his home in Mexico. Workers have not, of course, actually disappeared, but through dispersal in networks, they are treated as nonpersons. When the sleep dealers collapsed from the rigors of their work, they received no medical treatment and were immediately replaced by other workers in need of jobs.

1.2 Automation

The idea of automation has a long history having to do with labor-saving mechanisms that relegate some of the tasks performed by human beings to machines. According to Beniger (1986, p. 295), the term was introduced in 1936 by General Motors executive Delmar Harder to refer to "the automatic handling of parts between progressive production processes." By 1948, the word automation was used in the popular press as a synonym for "automatic control" (ibid.). Beniger traces the history of automatic control as far back as the third century B.C., when the waterlock represented the first instance of feedback control. In modern times, innovations such as the fantail to control the direction of a windmill (1746) and the centrifugal pendulum to control its speed (1787), the thermostat to control temperature (1830), the pneumatic proportional controller (late 1920s), the colorscope for precise color identification (1930), and pneumatic transmitters for industrial process control (mid-1930s) provide examples of early mechanisms of automation, with the Jacquard loom the first fully realized mechanism of digital control. These innovations promised a "control revolution" that transformed "an entire social processing system, from extraction and production to distribution and consumption" (Beniger 1986, p. 219).

Although automatic control could be implemented through any number of mechanisms (mechanical, hydraulic, pneumatic, electronic), wide-ranging automation with the goal of relegating human labor to machines came about only with the advent of modern digital computers in the years after World War II. Computer scientist Michael Dertouzos (1979, p. 38) defined automation as "the process of replacing human tasks by machines' functions." The process applied not only to manual labor, but was intended to incorporate significant elements of human cognitive tasks such as data processing, decision making, and organizational management. This goal was pursued through AI projects that sought to mimic and implement the most abstract forms of human thinking, such as solving mathematical problems, proving geometric theorems, and playing chess. There were early eye-catching successes in some areas, generating great enthusiasm for the prospects of computing and leading AI advocates such as Herbert Simon to the startling conclusion that "man's comparative advantage in energy production has been greatly reduced in most situations—to the point where he is no longer a significant source of power in our economy" (1960, p. 31). The logic of the capitalist economy, however, dictated otherwise, as we will discuss.

1.3 Augmentation

While control through automation remained a staple of the capitalist system, the obstacles encountered in delivering full automation needed to be overcome. This goal can, in principle, be accomplished through an alternative approach that was referred to as "intelligence amplification." Through this approach, "intellectual power" could be amplified through the introduction of "appropriate selection" rules (Ashby 1956, p. 272). This idea was taken to its logical conclusion by figures such as J. C. R. Licklider and Douglas Engelbart through the concepts of "man-computer symbiosis" and "augmenting human intellect." Licklider (1960) envisioned a future where "human brains and computing machines will be coupled together very tightly, and…the resulting partnership will think as no human brain has ever thought." The goal was to help human beings in "finding solutions to problems that before seemed insolvable, [including] the professional problems of diplomats, executives, social scientists, life scientists, physical scientists, attorneys, designers—whether the problem situation exists for twenty minutes or twenty years" (Engelbart 1962). The dream, in other words, was that computers would support and amplify human capacities in almost all areas of professional activity.

Augmentation, however, did not really come to life until a couple of long-haired college dropouts saw the potential for ordinary people to augment their activities with their own personal computers. The emergence of personal computing changed perceptions and visions with regard to the division of labor between humans and machines. It was during this period that Simon (1987, p. 12) observed that, "The almost simultaneous appearance of microcomputer and personal computers greatly broadened public awareness of computers in general …" Computers were ready to move into everyday life.

1.3.1 Personal Computers: Augmenting the Individual "User"
And thus it was that, not so very long ago, Steve Jobs and Steve Wozniak labored in a suburban garage to create the first marketable personal computer. The Apple I (priced at $666.66) was limited in capacity, but a huge breakthrough in making computing appealing to ordinary people. Wozniak had tried to interest Hewlett-Packard in the device, but the company[1] turned it down, believing that few people would want to own their own computer. Built of the cheapest parts Wozniak could find, and as few as possible, the Apple I fueled the primitive incubator for the future Apple Computer (now Apple, Inc.).

After selling a few hundred Apple I's, Wozniak began designing the much-improved Apple II. With this machine Apple Computer launched its business—and one of the most remarkable success stories in all of capitalism. The famous 1984 Apple television commercial, aired during the Super Bowl, linked the release of the machine with individual freedom and liberty.[2] Millions of Apple IIs were sold, with some remaining in use in school systems into the twenty-first century. The incredible utility of these computers is epitomized in the iconic program VisiCalc, the first graphical electronic spreadsheet. Dan Bricklin and Bob Frankston created VisiCalc on an Apple II. Bricklin said he dreamed up the idea while sitting in an MBA class at Harvard.[3] VisiCalc was modeled on the tabular grid of accountants' columnar paper, which contains numbered rows and columns. Today's spreadsheets, while enhanced in function, have not changed the basic VisiCalc format very much—it was that good. John-Louis Gassé, formerly of Apple, noted: "Approximations, trial and error, simulations—Visicalc is intellectual modeling clay. It lets you program without knowing it" (1987, p. 45). The machine replaced the error-prone human labor of calculation and displayed generous amounts of data with beautiful clarity. Humans figured out the logic of the problems they were addressing, programmed it into formulas, and assessed the results to make sure the algorithmic models were doing the right thing. Such unprecedented computational power in the hands of an accountant or small business owner in the late 1970s was deeply empowering (Nardi and Miller 1991). Apple transformed personal computers from hobbyists' toys into indispensable tools, building a company that, at various moments, has had the highest market valuation in the world.

VisiCalc was soon joined by applications for word processing, drawing, slide making, and gaming. These products drove the entry of personal computers into people's homes and lives. The applications made what people were already doing easier, as well as allowing people to do things they had never done before, such as playing imaginative computer games, drawing professional-looking illustrations with paint programs, and creating long newsletters with word processors. The first infamous family-news Christmas card newsletters appeared at this time! Other manufacturers, such as Radio Shack and Commodore, also produced personal computers. Human and machine were in sync, dividing complex cognitive labor between them, and reinforcing one another.

1.3.2   Workstations: Augmenting the Individual Worker
In the early- to mid-1980s, workers in advanced companies and organizations were supplied with computers that were then called workstations—innovative technology that vastly increased what employees could accomplish. A flurry of new machines—Xerox computers dubbed "D-Machines" (whimsically named after things beginning with *d*, such as Dandelion and Dorado), LMI Lambdas, Symbolics machines, and Sun Microsystems products—sparked a new era of workplace computing. These computers featured the new graphical user interface, whose design was based on the Xerox Alto, a prototype that had been around in small numbers since the mid-1970s (Campbell-Kelly and Aspray 1996). Workstations sported bit-mapped screens, point-and-click interfaces, and object-oriented languages. More powerful than personal computers of the era, these machines were *networked*. Applications such as email, bulletin boards, file sharing, chat, and group calendars

connected workers in new ways. (Such applications had been available before as Unix utilities and, in other forms, for technical workers in companies such as AT&T, and in technical departments in universities and think tanks.) Corporate work took on a more intimate, collaborative tenor. Communication was fast and often fun (the use of amusing typographic emoticons and jokey exchanges in email lightened the atmosphere). Work could be shared, critiqued, monitored, tracked, scrutinized, and organized rapidly and efficiently. Workers became informated, connected, sped up, digitally enhanced versions of their former selves. Little wonder that coffee and Mountain Dew (slightly in advance of Red Bull) were ubiquitous workplace drinks in offices equipped with the new digital technologies.

For the most part, workers embraced the changing practices digital technologies enabled. In addition to increasing productivity, responsibility, and participation at work, it was possible to develop wider networks of colleagues and friends, with whom it was now easy to stay in touch. Employees worked hard at work (and sometimes at home) because the technology allowed them to keep going even if all they had was a terminal in the home office. The computing subdisciplines of human–computer interaction and computer-supported collaborative work, both intended to enhance workplace productivity through better user interface and application design, were established (Grudin 2012).

Digital technologies were intensifying life with new connectivity, productivity, and a sense that anything was possible. This time was the era of AI startups, the dawn of open source software with its promise of sharing—a direct challenge to the commerce-oriented world of proprietary products—and widespread distribution of well-designed software tools such as Emacs and LaTeX. Text-based virtual worlds called MUDs and MOOs (the precursors of worlds such as Second Life), attracted passionate participants (Schiano 1999; Rheingold 2000). Continuing technical developments in personal computing and associated components and devices (such as chips and printers) in the hands of Microsoft, IBM, Intel, Apple, and Hewlett Packard, steadily increased computing power (Campbell-Kelly and Aspray 1996).

Still, though, in the mid- to late 1980s, the digitization of everyday life was tempered. Most networked computers were "mainframes, minicomputers, and professional workstations found at government offices, universities, and computer science research centers" (Zittrain 2008, p. 36). Though the distribution of powerful personal computing machines was changing rapidly as the 1980s drew to a close, most people didn't have cell phones. The Apple Newton, prefiguring what was to come in handheld computational devices, lay in the future. Most important, the World Wide Web had not yet arrived.

But change was afoot. Increasing intensity at work was felt with greater "demands for attention, vigilance, availability, and concentration" (Gollac and Volkoff, 1996; cited in Boltanski and Chiapello 2005, p. 247). If workers had powerful digital tools, it was expected that they would be more productive. The tools supported higher productivity, but not without workers adjusting their habits to take full advantage of the opportunities. Adjustments entailed a doubling down in which increased focus and concentration, as Gollac and Volkoff noted, were required to stay competitive. Those willing and able to learn the new technologies began to raise the stakes in terms of availability to customers, hours worked, and independence from corporate micromanagement. The augmented corporate worker was more independent, self-guided, focused, responsible, and assertive than ever.

These qualities fueled not only enhanced productivity in organizations, but the possibility of a contingent labor force of highly skilled, independent contract workers. Although free agency is a 1990s meme, workers in the 1980s began to perceive the erosion of secure lifetime employment. Some felt they might make an end run around the situation by working for themselves (Kunda, Barley, and Evans 2002; Nardi, Whittaker, and Schwarz 2002). Boltanski and Chiapello (2005) observe that such arrangements allowed employers to offload costs of training, vacation, and sick leave onto workers. Those experiencing layoffs had to retrain and fend for themselves during periods of unemployment. These trends have intensified to this date. Contingent labor has continued to grow, as have cycles of what Silicon Valley workers used to call binge-and-purge employment practices characterized by layoffs followed by waves of rehiring. Today, offshoring (made possible by digital technology) has been added to the mix, but it was less prevalent in the 1980s for professional jobs.

The buy-in of technically savvy, augmented workers is one means by which the alteration of the white-collar employment contract, increasingly characterized by discrete, short-term obligations in a climate of employment insecurity, has weathered its storms. The digitally empowered worker was reassigned from the dependent employee of a paternalistic corporation to an independent, savvy individual capable of taking care of herself, without expectation of many of the workplace protections formerly in place (Boltanski and Chiapello 2005). Today's millennial workers—self-sufficient, capable of long hours of work as long as they set them, restless, easily bored, heavily networked, creative, confident, immersed in digital technology—are the current manifestation of the turbulent forces set in motion during technological and economic shifts in the 1980s.

1.3.3 The World Wide Web: Augmenting the Masses

The workstations of the 1980s, initially designed and deployed by advanced knowledge workers in the tiny geography of Silicon Valley, merged, technically and economically, with personal computers. By the end of the 1980s, into the early 1990s, many white-collar workers performed their daily tasks with powerful, networked personal computers (Campbell-Kelly et al. 2013). The stage was set for the World Wide Web. When it arrived in a big way with the release of the Mosaic browser around 1993 (ibid., p. 289), the rapid uptake of Web services seemed to change the world overnight. People shifted from shopping at brick-and-mortar stores, calling companies on the phone, going into the bank, picking up airline tickets at local airline offices, and reading paper newspapers, to a world in which nearly everything was on the Web. The texture of everyday life altered, and "life on the screen" came online (Turkle 1997). E-commerce established itself, with, for example, Amazon.com going live in 1995 (selling books only). Electronic banking transformed managing everyday finances, and it became common to file taxes electronically (though this had been possible since the mid-1980s). Video gaming flourished. New genres emerged or were enhanced, including first-person shooter, real-time strategy, survival horror, and massively multiplayer online games. Open source software continued to develop, aided by enhanced networking. In the academy, the Association of Internet Researchers (AoIR) was founded in 1999 to keep up with research on the changes the World Wide Web had brought.

1.4 Heteromation

Heteromation gave rise to a new sociotechnical configuration with its own set of technologies, organizational arrangements, requisite roles and skills, and division of labor between humans and machines. Today, we observe the widespread use of heteromated labor in which the human operates on the margins of machines and computerized organizations. While automated systems relieve humans of labor, heteromated systems demand it. With the increase in heteromated systems, the large population of human beings who had been driven out of computing through automation, or never engaged with computing at all, are drawn back into the computational fold in new ways (Ekbia and Nardi 2014).

Although automation appears to be taking away jobs, "jobs" are in fact changing. Marx's core observation about the necessity of labor to capital accumulation continues to hold. Heteromation is consistent with the labor theory of value; capital uses computing to extract low-cost or no-cost labor in networks to sustain the growth of profits. People are not sitting idle, without work; we are working in new ways—ways that sometimes exacerbate precarity of employment and the safety net of workers' compensation, disability insurance, and old-age security. However, the story is rich and complicated, and, as we will explore, some of the labor we provide to profit-making organizations is freely chosen and very enjoyable; it is far from "alienated" in Marx's sense. Looking ahead, though, heteromated labor may be contributing to changing the employment relation and the nature of the economy to a system of tiny moments of economically valuable labor that return little to the worker but sustain wealthy, powerful companies.

1.4.1 Technologies of Heteromation

Developments in automation and augmentation were shaped by the historical circumstances of their respective eras, including the socioeconomics of capitalism. Heteromation is no different, and the division of labor between humans and machines leverages our histories as individuals in society and our advances in producing new technologies. In dividing labor between humans and machines, heteromation depends on both uniquely human social capacities that lie beyond what machines are capable of, as well as increasingly "smart" machines whose capacities continually expand. Human capacities change more slowly than technology—the development of a person as an emotional, cognizing subject is a lifelong project, whereas a disruptive technology can appear practically overnight. Heteromation leverages the development of persons, as well as digital technologies.

What can we say is different about machines vs. people, without essentializing their respective attributes? A clear boundary seems to demarcate the organic human capacity to care about others, ourselves, our survival, and our projects, from the indifference of machines. The caring capacity is ubiquitously manifest in networks through the participation of the masses in social media, gaming, citizen science, and much more, as we will discuss. On the other hand, a shifting boundary characterizes the cognitive capacities of humans relative to machines. We commonly think of humans as inherently good at certain kinds of cognitive tasks that computers perform clumsily or not at all. But the list of things that machines are good at constantly grows, and assertions about what humans are said to be good at generally consider only whether a human can accomplish a task, rather than whether the task is morally and ethically defensible or desirable. Since the cognitive boundary is fluid, capitalists' efforts to enlist humans to perform cognitive work depend on whether it is cheaper to hire a human or to program a computer. The question "to automate or not to automate" is ultimately an economic one.

Our perspective counters common assumptions in the literature—for example, Langlois's claim that "the quite different cognitive structures of humans and machines (including computers) ... explain and predict the tasks to which each will be most suited" (2003, p. 167). Drawing on evolutionary psychology, Langlois refers to "the kinds of cognition for which humans have been equipped by biological evolution." At the same time, he declares that "the human machine [sic] is also in many respects a response to economic forces. The human brain (and the human being more generally) is an evolved product, and therefore one whose design reflects tradeoffs motivated by resource constraints" (2003, p. 178).

We concur that the human condition is largely shaped by economic forces. Unlike Langlois, who views this reality from an essentialist evolutionary perspective, we see the development of humans and technologies as the outcome of specific, historical, socioeconomic changes. We reject innatist assumptions about cognition, and therefore, in subsequent chapters, we question notions about what is "suitable" for humans by contextualizing empirical examples within their particular historical conditions. Assuming a timeless, natural division of labor in which work is divided for humans and for machines, each according to putative cognitive abilities, deflects attention from the specific arrangements under which humans labor and the changing systems of compensation and reward in which they contribute value to the projects of others. This perspective also fails to acknowledge the steady technical progress humans have made in programming machines to perform increasingly sophisticated cognitive work.

1.4.2 Heteromated Labor and Its Value

Computer technology plays the dual role of *securing* value and *hiding* it at the same time. Burawoy (1979, p. 261) observed that capitalism is faced with the dilemma to "secure surplus value while at the same time keeping it hidden." This dilemma sets capitalism apart from social formations of the past such as slavery and feudalism, where the extraction of value from labor was explicit and transparent. Slavery extracted value through the direct ownership of the human laborer. Feudalism extracted value through separation of the work of subsistence that serfs performed for their own survival, and the work they had to do, without compensation, for the lord of the manor. In these two systems, securing value was made possible through extra-economic mechanisms of force and coercion.

In capitalism, by contrast, the extraction of labor is hidden from the laborer (and often from capitalists themselves), and the worker does not work by direct coercion, but from the need to earn a living to purchase food and shelter in the absence of access to means of subsistence production. The hidden character of value extraction goes hand in hand with the "voluntary" character of participation, giving capitalist control a whole new form (Marx 1990, I, chap. 7, 283–306).

Heteromation is a labor relation, and a quintessentially capitalist one in which economic value for others is extracted through labor. Heteromation is not an engineering concept or a matter of human–computer interaction. It does not refer to a particular set of tasks or actions, such as crowdsourcing, and it is not captured in terms such as human computation. Heteromation *always* extracts economic value for others. Crowdsourcing, for example, can be used to support the internal interests of a group or community. One system allows conference organizers to crowdsource participants' opinions about how to organize papers in sessions. The information is used to make the conference more enjoyable and reflective of participants' interests and preferences (André et al. 2013). All the value goes to the community.

Heteromation generally involves hidden labor, often unbeknownst to participants themselves, who are unaware that, for example, their social media activity may be transformed into commodities sold for advertising (van Dijk and Nieborg 2009, Ritzer and Jurgenson 2010), that self-service systems save corporations money by eliminating paid workers, or that production of gaming guides means that game companies do not have to hire technical writers. Revealing this hidden labor is all the more critical given the broad and expanding range of such contributions in the heavily computerized environments of contemporary societies.

1.4.3 Heteromation at a Glance

Our first attempt to theorize heteromation centered on what we called "inverse instrumentality." We examined systems such as video gaming where "subjects are bound into the system as necessary functional components"—that is, the system requires human intervention if it is to work at all (Ekbia and Nardi 2012, p. 169). We conceptualized inverse instrumentality as a mode of objectification and identified two groups of technologies that either *fragment* human individuals from within (e.g., personal health records) or *totalize* them as subjects within a totalized logic of engagement (e.g., computer games). Individuals, largely left to their own devices, are expected to "manage" their health, but also their health information. Complex video games cannot be played without the support of players who train and discipline other players (Nardi 2010; Ekbia and Nardi 2014; Kou and Nardi 2014). Since this earlier work, we have encountered a large and growing set of cases that can be more usefully understood as different types of heteromation. Part II of the book is dedicated to a description of these types, illustrated through specific cases.

It might be difficult, given the variety of types of heteromated labor spreading into so many aspects of contemporary life, to grasp the big picture. We hope it will be useful, therefore, to glimpse this picture before we embark on the details in the coming chapters. To do that, we present a contrastive image of heteromation to a vision that was put forth by one of the most influential figures of the twentieth century—Herbert Simon.

In the early days of computing, Simon, whose work spanned economics, psychology, computer science, artificial intelligence, management, and decision science, envisioned the following:

> Within the very near future—much less than twenty-five years—we shall have the technical capability of substituting machines for any and all human functions in organizations. (1960, p. 22)

More than a decade later, in the introduction to *The New Science of Management Decision*, Simon reiterated the vision:

> that computers and automation will contribute to a continuing, but not greatly accelerated, rise in productivity, that full employment will be maintained in the face of that rise, and that mankind will not find life of production and consumption in a more automated world greatly different from what it has been in the past. (Simon 1977, pp. 6–7)

This vision was based on the conviction that "in our times computers will be able to perform any cognitive task that a person can perform" (ibid.). Being an economist, Simon was aware that what will ultimately determine the relationship between humans and computers is not their respective technical capacities, but the socioeconomic system that embeds and enables those capacities. Accordingly, Simon discerned three dimensions of possibility on the impact of computer technology on the processes of management: a technological dimension, a philosophical dimension, and a socioeconomic dimension. According to Simon, expert opinions varied according to how small or large an impact was envisioned. Aligning the first two dimensions, Simon recognized four possible schools of thought in this respect (table 1.1). He characterized his own position as "fairly extreme along all dimensions": as a technological radical, an economic conservative, and a philosophic pragmatist (1977, p. 6). Simon maintained this position with typical intellectual doggedness throughout his career, facts on the ground notwithstanding.

Simon's vision, self-characterized as technologically radical but socioeconomically conservative, does not seem to have materialized many years and decades after its proclaimed forecast. What has happened instead is that humans are crowded into tasks that put them at the service

Table 1.1

Simon's classification of views on the impact of computers on management

	Socioeconomic	
	Conservative	*Radical*
Radical	Computers are limited in power, and business is done as usual.	Computers are limited in power, but there will be plenty of goods and services.
Philosophical/ conservative	Computers equal humans in terms of capabilities, but business is done as usual.	Computers equal humans in terms of capabilities and will replace humans.

of machines and the organizations that own and operate them, inverting both heads of Simon's vision.

We contend that the division of labor between humans and machines is sometimes, but not always, based on what each is good at. Langlois (2003), following Simon, spoke of "comparative cognitive advantage." In our view, the comparative advantages of humans are not so much in terms of cognition—which computers are increasingly good at—but in communicative, creative, and social activity. The social robot PARO, for example, provides stimulation and interest for people with dementia. But it does not function without the human mediation of engaged, caring workers and family (Chang, Šabanović, and Huber 2014; Ekbia, Nardi, and Šabanović 2015; see also chapter 8).

Cognitive labor is subject to shifting between humans and computers as computers become more capable. For example, currently the crowd-work platform TextBroker enlists workers to write simple texts for minimum wage or a little above. But machines are learning to generate such texts pretty well, using human-created textual corpora (Podolny 2015).[4] This development could push TextBroker into using humans to provide the creative labor of specifying story ideas, assigning capable machines to do the rest.

Creative labor is human for many reasons, including that humans care about things, as we have noted (see Heidegger 1962, Leontiev 1978, Emirbayer and Mische 1998, Kaptelinin and Nardi 2006). Heidegger (1977, p. 73) said, "Anything that is alive is bound together by the two fundamental tendencies of enhancement and preservation, i.e., 'a complex form of life.'" Spurred by plans and projects for enhancement, or by survival needs, humans create. Caring drives us to participate in the economy through our biological, emotional, social, and cognitive capacities as living beings.

Machines do not exhibit such participation, and this means that they have no need for social media or for shopping on Amazon.com, or any of the multitude of socioeconomic behaviors that humans exhibit. In this regard, machines are of no interest to capital. By themselves, machines could not sustain an economy (ambitions of pure automation notwithstanding). In a very deep sense, machines do nothing of their own accord until we activate them—prior to which we have dreamed them up, designed them, and built them. This applies as much to automated systems of the past as to the heteromated systems of today. As beings with objectives, or "object-orientedness," as cultural historical activity theory says (Vygotsky 1986), we become interesting to the economy as workers and customers. In the case of heteromation, the labor includes our work as generators of streams of data valuable to advertisers, employers, insurance companies, healthcare providers, planners, governments, and, sometimes, less reputable actors, such as those who hack into our accounts or steal our identities.

Some heteromated labor is constituted by the availability of an army of unemployed, part-time, or homebound laborers who seek extra sources of income, performing tasks machines could perform with sufficient (expensive) programming, such as image recognition tasks in Amazon Mechanical Turk. Other heteromation, such as free self-service labor, is based on considerations of efficiency and profit for corporations such as retailers, banks, and insurance companies, not on the quality of the service provided. The free labor in activities like citizen science and social media is stimulated by affective rewards such as the microvalidation of social connection and feelings of being useful and engaged. In part II of the book we discuss these forms of labor, and others, in detail.

1.4.5 From Automation to Heteromation: The Economics of Computing History

The brief history outlined here provides a sketch of the entwined history of computing and socioeconomic developments of the last few decades.

In particular, it shows how technological innovations have shifted the labor landscape for human beings, redefining the mechanisms of value extraction from this labor at each step. Rather than a linear process of technical development and progress, these shifts have followed a dialectic process wherein technical innovation responds to socioeconomic change and, in turn, acts as a driver of this change in a helical pattern. In this fashion, innovations of earlier stages are subsumed in more recent developments, often in response to the dictates of capital growth. Augmentation technologies subsumed automation and are in turn subsumed under heteromation in the next stage. The data processing and networking infrastructures that automated accounting, banking, and financial transactions, for instance, relieved humans of tedious and repetitive calculations, but they also save the banking industry in wages, benefits, and other labor costs. These technologies and infrastructures, in turn, enabled the introduction of ATMs, online banking, and more recently, self-service banking that puts a large burden on customers (as we discuss in chapter 6).

In summary, heteromation can be understood as a computer-mediated mechanism of extraction of economic value from various forms of human labor through an inclusionary logic, active engagement, and invisible control. These key attributes of heteromation—*inclusion, engagement, and invisibility*—make it at once novel, powerful, and dangerous.

2

The Logic of Wealth Accumulation: A Story of Political Economy

The brief, revisionary history of computing presented in the previous chapter makes better sense if embedded within the broader history of modernity. That history can be, in turn, written from different perspectives: social, political, cultural, economic, institutional. Daunting accounts by some of the great thinkers of the twentieth century, who sought to provide an inclusive image by bringing these perspectives together, create an image of modernity driven by incompatible logics and imbued with tensions, paradoxes, and disruptions (e.g., Weber 1904, Polanyi 1949, Heidegger 1962, Arendt 1958, Foucault 1977). All these accounts inform the discussion of computing and economy presented here, but we have found the perspective of political economy to be most aligned with our intellectual pursuits (Ekbia and Nardi 2015). With the dominant socio-economic system of modernity—capitalism—as the focus, views on political economy have undergone numerous changes over the decades, giving rise to a tumultuous and interesting intellectual history.

This chapter provides a brief overview of this history, as embodied in the works of observers and commentators of capitalism, past and present, with particular attention to the question of the accumulation of wealth in capitalism. Our aim in providing this overview is to acknowledge our intellectual debt to these thinkers, to show our lines of convergence and divergence with them, and to lay the conceptual groundwork for the remaining chapters of the book. Given our interest in computing, we spend more time on accounts of contemporary capitalism.

2.1 The Classic View: Accumulation through Innovation

The industrial revolution of the eighteenth and nineteenth centuries, which turned agricultural economies into capitalist ones, introduced social, moral, and political questions about the division of labor in society, the production and distribution of wealth and resources, and the governance of the economy. These questions engaged leading figures such as Adam Smith, David Ricardo, and John Stuart Mill as economists and moral philosophers of the time.

The opening sentence of Adam Smith's *Inquiry into the Nature and Causes of the Wealth of Nations* can, in fact, be considered the founding statement of modern political economy:

> The annual labour of every nation is the fund which originally supplies it with all the necessaries and conveniences of life which it annually consumes, and which consist always either in the immediate produce of that labour, or in what is purchased with the produce from other nations. (Smith 1776, p. 10)

The proportion of people who benefit from this national labor, Smith argued, is determined, first, "by the skill, dexterity, and judgment with which its labour is generally applied," and, second, "by the proportion between the number of those who are employed in useful labour, and that of those who are not so employed" (ibid.). Giving more significance to the first parameter, he then goes on to show how "civilized and thriving nations" have managed to accomplish abundance, "though a great number of people do not labor at all." This feat is opposed to "the savage nations of hunters and fishers," who cannot provide their people with the basic means of subsistence, despite the fact that everyone "is more or less employed in useful labor" (ibid.).

In this fashion, celebrating the productive power of labor in modern societies, Smith attributes it to "the quantity of capital stock which is employed" in setting labor to work (ibid., p. 11). A good part of this capital is invested in the supply of machinery that allows, through the proper division of labor, a modern pinmaker to produce almost five thousand pins a day—a revolutionary improvement compared to the production of one pin per day by the primitive pinmaker. Technology, therefore, had to also be celebrated if we note "how much labor is facilitated and abridged by the proper machinery" (ibid., p. 19). Much of this facilitation, in fact, derives from the laborers themselves who, now being focused on one single task, "are much more likely to discover easier and readier methods of attaining any object"—the playful boy, for instance, who reduced part of his labor by attaching a string to a fire engine valve (ibid., p. 20). High labor productivity, enabled by capital and driven by technical division of labor, was, in sum, Smith's explanation for the affluence of capitalist economies.

2.2 Marxism: Accumulation through the Extraction of Surplus Value

Karl Marx begged to differ. He agreed with Smith on the idea of human labor as the source of all wealth, and even adopted his theory of "original accumulation," but he was adamantly opposed to the idea that the accumulated wealth benefits all sides equally. He detected antagonistic interests where others saw harmony, recognized poverty and alienation when others spoke of progress, and counseled revolution where they prescribed hard work:

> *The worker need not necessarily gain when the capitalist does, but he necessarily loses when the latter loses* ... The political economist [Adam Smith] tells us that everything is bought with labor and that capital is nothing but accumulated labor; but at the same time he tells us that the worker, far from being able to buy everything, must sell himself and his family ... Whilst the interest of the worker, according to the political economists, never stands opposed to the interest of the society, society always and necessarily stands opposed to the interest of the worker. (Marx 1844, emphasis in original)

Marx traced the origins of his disagreement with the dominant political economy of his time in the observation that "it takes the interests of the capitalists to be the ultimate cause, i.e., it takes for granted what it is supposed to explain" (ibid.). To correct this problem, Marx proposed to proceed from "*actual* economic fact," rather than from the "primordial condition [that] explains nothing" (ibid.). He sought, in other words, to answer the questions and concepts of political economy—labor, wage, value, profit, property, money, capital, land, rent, production, exchange, consumption—by examining the dynamics of socioeconomic change, which he considered to be historical and dialectic rather than natural and objective. Conceptualizing production and consumption as an integrated whole, for instance, Marx posited, "Production thus produces not only the object, but also the manner of consumption, not only objectively but also subjectively. Production thus creates the consumer ... [It] not only supplies a material for the need, but it also supplies a need for the material ... [It] produces the object of consumption, the manner of consumption and the motive of consumption" (Marx 1939/1993, p. 92). Applied to current circumstances, this insight invites us to critically examine a host of issues—for example, the common notion that the adoption of digital technologies is driven by "need."

Marx's explanation for capital accumulation is the most celebrated example of his approach. The mechanism of surplus value creation, which revealed what he aptly characterized as the "mystery" of the reproduction of capital, is the distinctive feature of the Marxist theory of capitalist exploitation that sets it apart from other class theories such as Max Weber's (Wright 2005, p. 23). Where Weber focused his class analysis on the mechanisms of market exchange and how they give rise to unequal "life chances," Marx identified a second (and more important) mechanism—surplus value creation in the process of production—to account for the accumulation of capital through the exploitation of labor. Marx's theory of value is also distinct from those of economists such as Ricardo, who associated the exchange value of commodities with the amount of individual labor time that they embody. Unlike these theories, Marx considered the abstract notion of necessary labor-time as the appropriate measure of exchange value because he believed that labor is a fundamentally social phenomenon that can only be defined within specific sociohistorical relations.

Therein lies the power and beauty of Marx's analysis which is not lost on those who seek to understand the logic of accumulation in late capitalism. David Harvey, for instance, through a critical engagement with Marx's theory, has developed the concept of "accumulation by dispossession" to show how, contrary to Marx's prediction, new forms of accumulation are still at work to violently or hideously take common assets away from whole populations and territories on a global scale. He argues that, "capitalism must perpetually have something 'outside of itself' in order to stabilize itself" (Harvey 1999, p. 140). In other words, within a continual expansionist paradigm of activity, capitalism constantly searches outside the boundaries of its current hegemonies for new resources, new labor, new customers. This phenomenon has taken any number of legal or illegal forms, such as privatization of land, water, and other natural resources; conversion of common property rights into private ones; suppression of indigenous forms of production and consumption, and predatory credit schemes such as were widely practiced in the run-up to the 2008 housing crisis in the United States and elsewhere. In a nice emulation of Marx's method, Harvey applied this "inside-outside" dialectic to examine the relation between "expanded reproduction [of capital] on the one hand and the often violent processes of dispossession on the other" (ibid., pp. 141–142).

One of the key questions that arises from these observations is the relation between accumulation through dispossession and accumulation through exploitation of waged labor. This question is central to current debates about the applicability of Marx's theory to current capitalism and, hence, relevant to heteromation: Where does heteromation fit into this scheme? We pick this question up at the end of this chapter, and pursue it further in the rest of the book.

2.3 Neoclassical Theories: Accumulation through Rational Choice

The moral and historical perspective of Marx's political economy was challenged by those who sought to build a science of economics on the model of hard (mathematical) sciences, giving birth to the academic discipline of economics that has come to be associated with neoclassicism (Marshall 1890). With a focus on commodity prices and how they are fixed through the equilibrium between supply and demand, neoclassical economics is based on a set of key assumptions about human behavior. Most remarkably, it takes humans as rational and autonomous actors who seek to maximize the utility of their choices on the basis of full and relevant information. The market, according to neoclassicists, is the perfect environment for this kind of behavior, providing the arena for the efficient exchange of goods, as well as the right information about them to enable rational choice. By providing informational feedback, they argue, the market can correct initially incorrect models held by individuals, punish deviant behavior, and lead the survivors to the correct model.

The origin of wealth in the neoclassicist perspective, therefore, is in the rational choices of interest-maximizing individuals and profit-maximizing firms. Other than the fundamental assumption of rational choice, standard neoclassical models of capital growth (e.g., the Böhm-Bawerk model) are based on a set of other simplifying assumptions that are hardly grounded in reality—such as that labor and capital are continuously substitutable without limit (Waterman n.d.). This is a point that has been made by economists of different stripes for a long time (e.g., Kaldor 1962). Neoclassicists' indifference to such concerns is justified by their quest for a "rigorous" science of economics. The accomplishments of classical physics in theorizing a frictionless world greatly appealed to neoclassicists, who celebrated the analogy between economics and physics as an illustration of the analytic power of "unrealistic" assumptions (Friedman 1953, pp. 16–19). Whereas physicists were reminded by their lab instruments of the pervasiveness of friction, however, "[neoclassical] economists did not have a corresponding appreciation for the costs of running the economic system (Williamson 1985, p. 19).

In their desire to create a rigorous and "objective" science of economics that would be indifferent to moral values and political interests, however, neoclassicists have, in fact, contributed to the establishment of a particular socioeconomic order. In this fashion, the discipline of economics has performatively created the markets that it posits as natural entities. Markets have come to seem so natural that we do not, as an everyday matter of course, call into question their ontological status (Mirowski 2002).

The narrow focus of neoclassicism on rational behavior has been, in turn, challenged by thinkers on different points of the intellectual spectrum, giving rise to a new era of political economy. On the right, neoliberal economists have expanded the classicist horizon to argue that markets provide the best way of organizing not only the economy, but human affairs in general—individual labor, health, security, or a "marketplace of ideas" (Buchanan 1999). On the left, neo-Marxists discuss the impact of finance capitalism and globalization on issues of class, labor, and social equity (Harvey 2010). In between, institutional (Galbraith 1985), cultural and feminist (Federici 1982, Huws 2003, Peterson 2005), and ecological (Foster 2002, Klein 2014) scholars and commentators have focused on issues such as power and influence, domestic labor, and the environment.

2.4 Institutional Theories: Accumulation through Information

Institutionalists, for instance, have drawn attention to issues of heterogeneity of interests, ideological constructs, cultural constraints, and power, showing how politics influences economic decisions and policies, typically giving them a suboptimal character. By highlighting the incomplete and asymmetric character of information available to various players, institutional economists draw our attention to the ways by which institutions define, alter, and constrain the set of choices that individuals have at their disposal.

Individuals act on incomplete information, and with subjectively derived models that are frequently erroneous; the information feedback is typically insufficient to correct these subjective models. Institutions are not necessarily, or even usually, created to be socially efficient; rather, they—or at least the formal rules—are created to serve the interests of those with the bargaining power to devise new rules (North 1990, p. 16).

The emphasis on information asymmetry here derives from a view of economic activity known as transaction cost economics. Broadly defined as "the costs of running the economic system" (Arrow 1969, p. 48), transaction costs are compared to friction in physical systems. In a celebrated essay, Ronald Coase showed that neoclassical theory is limited in its application only to zero transaction cost conditions—that is, to those situations where information and transactions have no cost. Information and transactions, however, have costs, which must be figured into the valuation of a good, above and beyond the costs of production (land, labor, machinery). These costs have to do with "defining, protecting, and enforcing the property rights ... [including] the right to use, the right to derive income from the use, the right to exclude, and the right to exchange" (North 1990, p. 28)—in short, the cost of measurement and enforcement.

Whoever defines these rights and rules, therefore, is in a better position to accumulate wealth through property protection, information hiding, deception, or any number of other devices. In this fashion, information asymmetry leads to income asymmetries, putting wealth in the hands of certain individuals, firms, communities, and nations at the expense of others. This is, roughly, the answer of institutional economics to what North considers the "central puzzle of human history"—namely, how societies have diverged in terms of their performance, wealth, and well-being.

2.5 The Sharing Economy: Accumulation through Networking

Coase's idea takes on a rather different reading by another group of writers who take their cues from the success of crowdsourced projects such as Linux and Wikipedia. These projects, driven, implemented, and maintained by large groups of volunteers who freely contribute their skill and knowledge without monetary reward, introduce a new model of production that is different from both markets and firms (Benkler 2002, 2003). Coase had shown that the choice between markets and firms as the organization of economic activity is based on their relative transaction cost, and institutional economists such as Williamson generalized this insight to an extended theory of organizations and institutions of capitalism. Cases such as open source software, however, do not seem to fit into this general institutional theory because apparently they neither have the pricing mechanism of markets (e.g., there is no wage) nor the hierarchical structure of firms (e.g., there are no managers).

This is the central argument of a number of writers, who have, in turn, generalized from the case of open source to a broader phenomenon that encompasses peer or social production (Benkler 2002, 2006), the zero marginal cost society (Rifkin 2014), open cooperativism (Conaty and Bollier 2014), the cognitive surplus (Shirky 2010), and human computation (Michelucci 2013). Rifkin, for example, argues that capitalism will shrink because people can produce their own information, print material goods on 3D printers, generate green electricity at home, and educate themselves through online courses. Soon goods and services will be "nearly free" (2014, p. 4).

There are noticeable differences in detail and perspective among these various views, but their shared premise is that computing technology has opened up a space for a new mode of economic activity and production that does not fit into the traditional capitalist modes of accumulation. Depending on their political orientation, different authors and commentators see an opportunity for parallel economic activities that might ultimately undermine capitalism (at least in its most explicit forms of exploitation), or simply for a new means of wealth accumulation. There is empirical evidence to support either scenario, so in a sense, the jury is still out as to what might unfold in the long run. But there are serious reasons to be skeptical of the more positive scenarios that suggest that capitalism will gracefully recede.

2.6 Digital Labor: Accumulation through Free Labor

Where Benkler, Rifkin, and Shirky see in current computing practices a potential for peer and community production, others recognize a potential for new modes of exploitation. Terranova (2000), building on media theorist Dallas Smythe's idea (1981) of "audience labor," used the notion of "free labor" to refer to this emerging phenomenon. Others have since employed this idea to illustrate the economic function of digital technologies in a new capitalism. Fuchs, for instance, follows the specific thread in Marx's writing that defines class on the basis of the process of appropriation of surplus value—that is, the value produced by labor in the process of production above and beyond the value it needs to create for its own sustenance in the form of a wage. Fuchs' argument is that the creation of value has shifted from paid labor to unpaid (free) labor:

> Users employ social media because they strive to a certain degree for achieving what Bourdieu ... terms social capital (the accumulation of social relations), cultural capital (the accumulation of qualification, education, knowledge) and symbolic capital (the accumulation of reputation). The time that users spend on commercial social media platforms for generating social, cultural, and symbolic capital is in the process of prosumer commodification transformed into economic capital. Labour time on commercial social media is the conversion of Bourdieuian social, cultural and symbolic capital into Marxian value and economic capital. (Fuchs 2012, p. 638)

On this basis, all contributors to social media belong to what Fuchs dubs the "exploited class." Finding the traditional (industrial) notion of "working class" inadequate for the contemporary situation, though, Fuchs (2012) expands the exploited class to include direct knowledge workers (in, e.g., health, education, and other service industries), indirect knowledge workers (e.g., homemakers, largely female, who "produce knowledge in the broad sense of communication, affects, sexuality, domestic goods and services"), and the "underclass" (e.g., the unemployed and underemployed, migrants, retirees), as well as the self-employed (ibid.).

Other observers, such as Galloway and Scholz, follow a similar tack, highlighting the erasure of the boundary between work and leisure as the mechanism that enables the monetization of "unwaged labor" on the internet. Wondering if the playful activity of updating one's "status" on Facebook can count as labor, Scholz (2013, p. 2), lands on the affirmative side, comparing the activity to "the invisible, unsung forms of traditional women's labor such as child care, housework, and surrogacy." Nonetheless, he cautions against the fetishization of computing that might distract us from the "real" places of exploitation—namely, the slums of economic developing countries: "Digital labor in the overdeveloped world is contingent upon the sweat of exploited labor in countries such as China" (ibid., p. 3).

2.7 Affective Capitalism: Accumulation through Persuasion

In the 1990s, a different line of thought on capitalist accumulation emerged in Italy under the influence of Michel Hardt and Antonio Negri's (2000) *Empire*. Variously called the "Autonomist" or "post-workerist" movement, this thinking is premised on the notion of "value-affect," which can be roughly understood in terms of the relationship between labor and affect, which Negri, following Spinoza, defines as "the power to act." The erasure of affect and subjectivity from the measurement of value, which is the linchpin of political economy, has generated an apparent paradox that Negri seeks to undo. The paradox derives from the fact that affect, as something that is not measureable, is at the same time at the center of value creation. To resolve the paradox, therefore, one needs to put affect back at the center of one's theory, where it belongs—hence, the notion of value-affect.

Negri's concept of value-affect provides the basis for the Autonomist view of current capitalism, which it says "takes the mind, language, and creativity as primary tools for the production of value" (Berardi 2009, p. 21). This concept also provides support to arguments against Fuchs's theory of value based on labor time. With labor becoming more complex (relying on affects, motivation, reputation), and value becoming more abstract and financialized (increasingly produced in complex networks involving firms and consumers but also financial analysts, brands, and so on), Autonomists argue that the economy has shifted toward an affective law of value, "where the values of companies and their intangible assets are set not in relation to an objective measurement, like labor time, but in relation to their ability to attract and aggregate various kinds of affective investments, like intersubjective judgments of their overall value or utility in terms of mediated forms of reputation" (Arvidsson and Colleoni 2012, p. 142).

A linguistically focused version of the Autonomist view is developed by Marazzi (2007, p. 21), who argues that communication has come to the fore in new modes of production that emerged in the 1980s, such as "lean," "just-in-time," and "post-Fordist" production:

... communication has become as important as electricity was in the age of mechanical production. In fact, communication is the grease that insures the smooth running of the entire production process, from the sale and distribution to the production stage.

Based on themes such as this, Autonomists advocate an alternative meaning of value and wealth "as the simple capacity to enjoy the world available in terms of time, concentration, and freedom" (Berardi 2009, p. 81), and find in digital technologies effective tools for the creation of this kind of wealth.

2.8 Digital Capitalism: Accumulation through Repression

The communication theorist and political economist Dan Schiller has a very different perspective on the role of information and communication technologies in current capitalism. In his earlier book *Digital Capitalism* (1999), Schiller chronicled the metamorphosis in the US economy "whereby capitalism was made over to accept a more ICT-intensive orientation" (2014, p. 5). Using examples such as Ford Motor Company, with its 120,000 workstations and vast intranet that connected factories and offices with 15,000 dealers, he shows the steep increase in capital investment in information technology by US corporations in the last quarter of the twentieth century, from 7 percent to around 45 percent. In this account, on a par with industrial capitalism, digital capitalism marks a new era in a five-hundred-year history with abiding tendencies:

> [We observe] capital's continually extended use of wage labor, its search for new and often contested sites of commodification, and its episodic crises, wherein rampant financial speculation triggers a fall into depression and economic stagnation. (Schiller 2014, p. 8)

In this way, Schiller expands his earlier work to demonstrate the role of digital technology in capitalist crises such as the recession of 2008. He traces out the origins of this recession in terms of three processes: (1) reorganization of the system of production, along with the restructuring of labor; (2) the infusion of capital into finance; and (3) escalating military procurement spending. His close examination of the history of General Motors in the last decades of the twentieth century sheds light on the first process. GM, faced with militancy from its factory workforce and competition from Japanese rivals, embarked on a multibillion project to bring information technology to the center of its production, bridging the "islands of automation" that were built earlier, and connecting more than 200 major facilities and 50,000 dealers, suppliers, and offices scattered around three dozen countries (Schiller 2014, pp. 34–35). The weakening of labor unions, later written into law by turning Michigan into a right-to-work state, was at the core of this process. Similar processes of heavy computerization in the financial sector and the military propelled the economic downturn of 2008 and the subsequent events that followed (ibid.).

The cumulative effect of these processes is what can be described as "accumulation through repression," as manifested most vividly in the growing security apparatus in cyberspace. Schiller (2014, p. 222) draws on a "green paper" by the US Department of Commerce to show how a major portion of the cost for protecting corporate networks fell on taxpayers, while at the same time subjecting ordinary citizens to the most severe forms of surveillance. In a similar vein, McChesney (2013) points to the internet's essential relationship with commercialism and advertising, especially the widespread collection of personal data, and a "corrupted" state commandeered by corporations who pursue our personal data, wresting it from our control.

2.9 Synopsis

This constellation of views shows the development of ideas about political economy, along with changes in the capitalist system in the last 200 years. The diversity of perspectives, particularly on contemporary capitalism, makes it difficult to draw any general conclusions, but a number of points can be highlighted. First, the growing consensus outside neoclassical economics is that modern social life cannot be understood in reductionist and mechanistic terms of "pure" economics because it is permeated with issues, questions, and predicaments that have a strong political economy character.

Second, the parallels between the development of political economic theories on the one hand, and the capitalist system as a socioeconomic arrangement with particular logics of accumulation, reward structures, and incentive mechanisms, on the other, illustrate the strongly "performative" character of economic thinking. Rather than a "natural phenomenon," the capitalist economy is, ultimately, an outcome of economic theory (Mirowski 2002). It is the economists who, in a performative sense, create the economy (MacKenzie n.d.). The importance of this point cannot be overemphasized in light of the commonplace misperception, propagated by mainstream economic historians (e.g., Langlois 2007), that the capitalist system and its course of development represent a "natural" state of affairs, deriving its character from, for instance, the competitiveness and rivalry "inherent" in humans as part of their biological evolution. This misperception accounts, in part, for the sense of helplessness and confusion prevalent in our culture today.

The third point to be extracted from this brief history is a growing understanding of the tight relationship between technology and economics, a relation that is perhaps underappreciated in both the popular imagination and academic and professional thinking—e.g., in the area of human–computer interaction, where the elephant of political economy is visibly present in the room, but rarely spoken of (Ekbia and Nardi 2015). As North (1990, p. 132) points out, with the exception of Karl Marx, "who attempted to integrate technological change with institutional change ... [technology] has essentially remained outside any formal body of theory" in economics, especially in the neoclassical framework. To the extent that technology is featured in economic thinking, it is usually portrayed as a source of wealth. Despite early economic historians such as Adam Smith,[1] who emphasized technology as the creator of human well-being, mainstream economics has remained largely silent on the economic predicaments generated by technology.

In summary, and in light of these observations, an analysis of computing from the perspective of political economy of the kind we pursue here can enhance our understanding of the relationship between computer technology and capitalist economies. Such a perspective seeks to understand computing practices within the purview of sociohistorical developments, socioeconomic systems, legal and regulatory frameworks, environmental impacts, governance structures, state power, and political agendas on local and global scales. The notion of heteromation provides a particular angle on this perspective, with a focus on labor and accumulation of wealth.

2.10 Accumulation through Heteromation

The account of heteromation presented in this book intersects with many of the ideas and views described here, aligning with some of them and diverging from others. Our contention, first and foremost, is that heteromation captures a new and expanding logic of wealth accumulation in contemporary capitalism, which overlaps with earlier logics but is distinct enough from them to deserve separate treatment. The idea of heteromation is consistent with the commodity chain approach, which, as Schiller observes, is rooted in the work of Wallerstein who examined flows of labor, consumption, and production in a growing global economy (1983). Schiller (2014, p. 7) notes that the *value chain* metaphor, deeply rooted in mainstream economic thinking since the 1980s, focuses on how capitalist *organizations* add economic value, sidestepping the wider *commodity chain* that emphasizes the value of *labor* in an "ever-reconfiguring process of capital accumulation." Heteromation is one form of computer-mediated labor in this contemporary context.

In particular, heteromation overlaps with Marxian theory in at least two respects: (1) class structure of capitalist economies; and (2) human labor as the source of value (Ekbia 2016). First, current capitalism, like all earlier forms of capitalism, is closely tied to a class-based society. *Class structures* are relatively stable, defined as they are by polarized relationships to means of production (e.g., capitalists and laborers). *Class formations*, on the other hand, are more dynamic, and depend on the ways collectivities organize themselves on the basis of their interests at any given historical moment (Wright 1997). The formations change according to the specific stage or "spirit" of capitalism, as well as the balance of social and political power in a given society. Class formation in American capitalism of the early nineteenth century largely consisted of family-owned enterprises and their employees; it shifted to large corporations controlled by non-owner managers in the second half of the century, and later to monopolistic cartels of the early twentieth century.

Class formations in contemporary capitalism have acquired a hyper-dynamic and fluid character, largely embodied in computer-mediated network relationships with a global span. Networks embody the class formations of contemporary capitalism. A key consequence of these changes is that current capitalism is "inclusionary" rather than exclusionary—that is, *it secures value by bringing and keeping large segments of the population into its fold in the form of unwaged, unpaid, or minimally compensated labor.* While, as Marx repeatedly pointed out, earlier eras of capitalism also benefited from the reserve army of labor available on the market, in the new world, where connecting to networks can be attained at very low cost, *exclusion would eliminate new means of value extraction.* Instead, digital inclusion—in the sense of being connected to a network, but *not* being a member of the privileged class—has become the *modus operandi* of current capitalism. Capitalism has reinvented itself once again, using the labor of the masses through digital inclusion. One can watch, in real time, as capitalism occupies itself with bringing more and more of the masses into computer-mediated networks—for example, Facebook and Google's plans to deploy drones to deliver their services to the farthest reaches of the globe.

Second, conventional waged labor has been, and remains, the main source of value creation in capitalist economies. Although the general principle has stayed constant—"to secure and obscure the extraction of surplus value" (Burawoy 1979, p. 254)—the techniques and mechanisms of extraction of value have changed throughout the eras. The historical trend in capitalism has been the employment of more indirect and diffuse forms of control that enable subtle forms of obscuring while expanding the circle of those whose labor is secured. Heteromation embodies this trend, revealing the mechanisms of value extraction despite their subtlety. Although it is increasingly ubiquitous, heteromated labor does *not* replace, nor does it contradict, the forms of value extraction that Marx analyzed on the basis of his theory of waged labor. To conflate the two is not only an analytic fallacy based on mistaking class formations for class structures, it would do injustice to both forms of value extraction. Heteromation and exploitation in the Marxian sense of the term represent two *distinct* forms of value extraction in current capitalism, and they should be understood as such. Heteromation involves free or minimally compensated labor, rather than waged labor with its traditional class relations of contractually bound workers and owners.

Heteromation is compatible with Harvey's notion of the "inside-outside" logic of capitalist expansion. If it is true that capitalism continually acts to appropriate what is "outside" its current activities in order to feed the process of accumulation necessary for its stability, then computing technologies have created a convenient vehicle for realizing this. These technologies are not only labor saving, as dominant accounts remind us—they are labor creating! While automation pushes human individuals outside the loop, heteromation brings them back in through various mechanisms of engagement, as we discuss at length (see, in particular, chapter 10). The dialectic of digital inclusion prevalent in the current networked world (Ekbia 2016), and enacted through heteromated systems, is one embodiment of the inside-outside logic.

We can say that heteromation takes up where accumulation by dispossession leaves off. Just as accumulation by dispossession is one mechanism of capitalist expansion, pushing capitalism toward territory "outside" current sources of value, heteromation is another. Both develop within an inside-outside logic, but heteromation is distinct in its implementation. It lacks the brutal, conspicuous mode of acquisition of accumulation by dispossession; in fact, heteromation succeeds by sneaking in on little cat feet, insinuating itself everywhere in computer-mediated networks through nearly imperceptible, dispersed, delicate methods of incitement. Heteromation extracts value through billions of tiny moments of labor in networks, rather than blatantly, visibly ripping away resources for capital as is typical of accumulation by dispossession.

How capitalism has arrived to its current state, how it has "invented" heteromation as a new form of value extraction, and how computing technology has enabled this transition is the story that we pick up in the next chapter.

3

The Dynamics of Capitalist Change: A Story of Resilience

3.1 Introduction

In this chapter, we recap historical developments that parallel, in some ways, the account of automation, augmentation, and heteromation in chapter 1, but focus on how, as individuals, we become entangled with technology in our everyday lives. Marx, Schumpeter, Mensch, Weber, and others have observed that capitalism is as much a social system as an economic system. Without question, capitalism is intimately bound up with how we narrate our life stories, interact with one another, and shape our projects and activities. Capitalism fundamentally depends on the labor in that socioeconomic system, in particular, the surplus value that can be extracted from labor.

Marx defined surplus value as the difference between a laborer's total labor-time (let us say, for a day) and the amount of labor needed to sustain the worker to ensure that he can eat, sleep, and thrive. The worker is paid for his subsistence, which Marx referred to as "necessary labor-time," while the remainder of the day's labor goes toward capital's profit as surplus labor-time converted into surplus value. Burawoy (1979, p. 260) explains that surplus value is "appropriated by the capitalist as unpaid or surplus labor time and later realized as profit through the sale of commodities on the market."

3.1.1 Surplus Value: Then and Now

In feudal times, workers provisioned themselves with food and shelter by working their own parcel of land part of the time. The rest of the time, they worked on the feudal lord's land, giving him the fruits of their labors (Burawoy 1979). Caffentzis (2005, p. 94), following Marx, notes that the serf was aware of exactly what he was giving up, while the waged worker in a capitalist system is not:

> [Under capitalism, surplus value] is formally and legally hidden by the wage form. As Marx frequently points out, it is clear to the serf when s/he is working on his/her land versus on the land of the lord, whereas for the waged worker the moment when the labor-time necessary to create the value of his/her wage is finished and surplus labor-time begins is systematically obscured by the wage form and the general process of valuation.

We are, of course, long past the days of serfs and lords, and to some extent, even beyond the mechanical production lines of the nineteenth century. The imposing factories of Marx's day have given way, in critical sectors of the economy, to work suffused at every step with digitally mediated, informated processes of production and distribution (Zuboff 1988). Why, then, is labor still the source of value? Why have we not progressed, with all our sophisticated machinery and technology, to something closer to self-reproducing automata that would relieve capital of the unruly, contentious, expensive persons who constitute the labor force?

Science fiction predicted such automata, or at least depicted them as one vision of the future. Philip K. Dick's story "Autofac" imagines an autonomous, machine-generated factory. After a devastating war, humans discover that the machines are restarting themselves underground. "'They're building,' O'Neill said, awed. The machinery was building a … replica of the demolished factory." "Autofac" was written in 1955. While plenty of automation has entered the economy since then, there are, in fact, no machines such as those Dick envisioned. Nor do any appear on the horizon. Marxian theory indicates that such automata will not appear in capitalism because machines, by themselves, do not produce value. Marx observed that machinery and paraphernalia, which he called "dead capital," are perpetually in need of "living labor" to animate and energize them.[1] Workers' living knowledge is deployed to imagine, design, and build machinery. And then to operate, maintain, repair, recycle, and finally, bury it.

We can imagine, and, to some extent, construct technologies such as software programs that learn, self-replicating machines that marshal resources for their reproduction, and so on—but it is always a human who imagines them and sets in motion the activities to produce them. The image of Dick's autonomous machines startles because the machines "care" enough to reproduce themselves as they once were. Their sense of history perturbs us. We are impelled to examine why we find the idea of caring, reflexive machines so fascinating, and, ultimately, so unreal.[2]

The autonomous machines of Dick's story would be suppressed by capital, were they ever to appear. Their agenda—to replicate themselves in their own historical image ("building a ... replica of the demolished factory")—is a reactionary dream incompatible with capital's forward-looking disruptive purposes. Marxist philosopher George Caffentzis observes that for capitalism, automation is at best a double-edged sword. On the one hand, it relieves capital of pesky workers, but on the other hand, since labor is the source of value, the rate of profit is in danger of falling. Caffentzis (2013, p. 72) remarks, "Hence, the capitalist class faces a permanent contradiction it must finesse: (a) the desire to eliminate recalcitrant, demanding workers from production, (b) the desire to exploit the largest mass of workers possible."

3.1.2 The Shell Game: Making Value Invisible

Heteromation nicely squares this circle. It engages millions of workers who do tiny jobs for little or no remuneration. The volume of work produced, along with no/low pay, help finesse the falling rate of profit by leveraging "the largest mass of workers possible." The widespread extraction of value from (mostly) uncompensated, computationally mediated labor is capitalism's new gig.[3] Obscuring many aspects of labor, heteromation's shell game hides surplus value under one of the shells. But, as in any well-played instance of the game, the shells are moving around so quickly we don't know which one has the surplus value. For example, marketing and product organizations may refer to Wikipedia articles in information products they prepare for customers, embedding Wikipedia links in the documents to inform customers about pertinent topics. In the old days, actual employees would have prepared such materials, which now come from the unpaid labor of Wikipedians.[4] Somewhere, under one of the shells, a Wikipedia author has been paid (at least) a subsistence wage so that she can work at unpaid labor appropriated by the marketer and product manager who sell Wikipedia-derived information products to customers. She could not perform the unpaid labor without the necessary labor-time to keep her alive. We don't know where that labor came from. Surplus value circulates in confusingly complex capital flows.

Maybe the Wikipedian herself is not the one actually getting the wage. She could be on welfare, or living in her (working) mom's basement, or supported by a friend or spouse. In any of the cases, she unwittingly contributes to the bottom line of the marketers and product managers who sell her intellectual work to their customers.

Across the board, digitally mediated work comes at low cost to capitalists. Amazon Mechanical Turk compensates workers, but only about half of the workers rely on their income for basic needs (daily expenses, education, credit card payments, other basic needs; see Jiang, Wang, and Nardi 2015), because it is really not enough to survive. Workers' subsistence comes from other necessary labor-time: from a job, or their families' necessary labor-time, or from taxes (i.e., contributions to AMT workers through welfare transfers). No one in the US can live on less than minimum wage with no benefits—which is what Mechanical Turk workers receive (Chilton et al. 2010)—so the necessary labor-time is made up from other sources.

The other half of the workers Jiang, Wang, and Nardi (2015) studied used the meager compensation for what they called "fun stuff" and small luxuries, gifts, electronics, and inessentials. Their subsistence was also coming from somewhere else, allowing Amazon to pay them a pittance, which they spent for nonessentials, including sometimes taking the wage in the form of scrip with which to purchase goods on Amazon.com, allowing the company to enterprisingly double dip by forging consumers of its own products out of AMT workers.

Further cost savings arise from the fact that online work requires little supervision. What is needed can often be handled algorithmically (Irani and Silberman 2013). Workers bring their tools to work; actually, they don't "go to work," so no offices or factories are necessary. Workers remain in their homes with their own computers and devices. There are no corporate expenses for janitors, HR managers, or IT departments. No benefits are paid.

Any child old enough to manage a mouse can begin to participate in heteromated labor. Indeed, a multitude of virtual worlds such as Club Penguin and Webkinz World ("recommended for ages 6 and up") draw children into computationally mediated entertainments, naturalizing the chatrooms and user-generated content that fuel advertising. Mobile phone platforms increasingly offer kid-friendly applications. Caffentzis (2013,

p. 73) suggests that the danger of a falling rate of profit could be mitigated by mechanisms such as "the reintroduction of slavery [or] a dramatic increase in the workday." He doesn't mention child labor, but remarkably, heteromation hides what is actually child labor, although in so benign a form we can scarcely use the freighted words to describe it. Heteromation is considerably easier to manage than bringing back slavery or attempting a longer workday—it leverages the zeitgeist of "freedom" of the neoliberal ideology, and the culture's attraction to technology, even at the tenderest ages.

3.1.3 Calculating Machines and Incalculable Labor

Machines, and, more broadly, any technology, will always be dead capital. But our relation to machines is complex because they powerfully attract us to set them in motion through living labor. The sources of this attraction are many, and have never been completely understood. Heteromation harnesses this attraction, but may also develop a coercive side. Whether voluntary or coerced, heteromated labor moves the human worker to the margins of the machine or organization. As Marx predicted, the worker "step[s] to the side of the production process instead of being its chief actor," and plays the role of "watchman and regulator to the production process" (Marx 1973, p. 705, quoted in Caffentzis 2013, p. 44).

Capital decides whether it makes a profit through its own methods of calculation. These methods comprise a social relation in which capitalists agree on what constitutes the "bottom line." But capitalism does not fully theorize itself, and its inputs are more complex than the quantities that appear on balance sheets. For example, feminist theorists of the 1970s and 1980s noted that labor power, i.e., the commodity sold to capitalists by workers, depends on other forms of labor, such as that of the household (Federici 1982). We have not devised ways of valuing such labor, and the capitalist social relation does not recognize it. Labor power is "a legal and economic category that is expressed and materialized through its institutionalization and sale through the wage labor contract," as Comor (2015, p. 14) points out.

Heteromation yields economic value, but we do not yet know how to measure this value since the convenient abstraction of a wage is not present (in most cases). As Andrejevic (2013, p. 151) put it plainly, "Although forms of online activity that take place outside the workplace proper do not fall within the realm of wage labor, they can nevertheless generate value." In capitalism, ultimately, value derives from waged labor, but within a process so circuitous it will have to be mapped by examining complex capital flows (Robinson 2015). Ekbia (2016) remarks that we need a "theory of value ... to identify the channels through which capital flows in order to maintain its growth."

Caffentzis (2013, p. 151) observes the integrated nature of the capitalist system as a whole (rather than autonomous individual capitalist units): "The capitalist who invests only in ... machinery, buildings, and raw materials and nothing in ... labor-power ..." is, nonetheless, embedded in a larger system in which value derives "from spheres of production that operate with [a great deal of labor]." The video game maker Blizzard Entertainment, for example, operates a vast network of servers but has fewer than 5,000 employees. Within this machine-rich, employee-poor configuration, Blizzard generates nearly $5 billion in annual revenue. However, the company is positioned inside an industry that relies on cheap global labor to produce its machinery, all the way down to the dangerous labor of mineral extraction and waste management required for the manufacture and disposal of computing machines. Blizzard's own products generate massive amounts of heteromated labor (as we will describe in part II of this book). The apparent magic of producing five billion dollars with five thousand employees is not the simple story it appears to be. Other big players in the software industry are, likewise, leveraging global systems of labor—heteromated labor as well as the physical labor of extraction, manufacture, and waste disposal of machines, much of it conducted in developing countries under low-wage conditions.

The parable of workers and serfs that opens this chapter reveals that capital extracts value for its profits from labor-time. But no available conceptual apparatus calculates the quantity of that value when there is no wage, as in much heteromated labor. This gap does not mean that the value does not exist. Caffentzis (2013, p. 132) remarks:

> Labor-power that has, or tends to have, zero value (i.e., it is wageless) can be enormously productive of surplus value through the total cycle of value production ... You can no more determine productive labor by paychecks than you can determine value by stopwatches.

He further notes that just because we cannot measure something now does not mean that we will not devise appropriate measurement in the future. We have, then, only partially glimpsed how capital extracts a cornucopia of economic value through heteromated labor within complex global systems. Many questions remain.

We now push back for some historical context, outlining the evolution of everyday technologies since the end of World War II. The goal of this discussion, in which we rather drastically compress decades of complex history, is a précis of economy and technology that considers persons wending their way through sometimes bewildering social change, engaging technology on the journey[5]. We do not aim to catalog anything or provide an exhaustive account, but we do want to shake the kaleidoscope to bring into view an image of how we have ended up with so many forms of heteromated labor. Over a period of decades, we have gradually intensified our engagements with technology and computing, providing capital the opportunity to extract value from our participation in networks. In subsequent chapters, we discuss forms of heteromated labor in detail, but for now, we pause to consider how, in the last 70 years, technology, culture, and society have co-developed in complicated and unexpected ways.

3.2 Modernist Tensions and the Paradox of Liberalism

The discussion begins with a contextualizing look at the paradox of liberalism. Historically, our time is the immediate heir to a socioeconomic system and form of governance called liberalism. Freedom is the purported historical gift of liberalism to humanity. The problem for the European societies in the eighteenth century (as for many societies of our era) was how to limit the exercise of power by a public authority, especially the person of the sovereign, and liberalism provided a solution to this problem (Foucault 2008). The solution basically consisted of the provision of an even playing field, a market, immune from state intervention (and, for that matter, from sovereign intervention), which gives everyone the opportunity to take part. The offering, in other words, was a fundamentally *economic* freedom. To this end, the liberal strategy was to decouple economy and social policy, not in the sense of severing their relationship, but in the sense of letting the market economy do its work in such a way that the social would be taken care of. Although there are various versions of liberalism in terms of sociopolitical setting, historical background, and underlying principles, they share this fundamental approach.

However, liberalism did not guarantee, provide, or even respect freedom, as attested by some of its most fervent advocates such as Milton Friedman (1962). What liberalism purported to do was to produce what we need to be free—the conditions, organizations, and instruments that create the possibility for the production of socioeconomic, legal, and political freedoms. Consequently, freedom in liberal societies had to be constantly produced and accomplished; it was not a given. This duality between freedom and the coercive instruments and techniques for producing freedom led to what Foucault called the "paradox of liberalism." On the one hand, the process of providing freedoms—individual, political, economic—has a light, positive, and productive side. On the other, it has a dark, negative, and constraining side. Capitalism has responded to the predicaments generated by its dark side through a double strategy of absorption and neutralization, with a tendency "to take back on one level what it offers on the other" (Boltanski and Chiapello 2005, p. 435). Capitalism finds openings for enriching itself in the very predicaments it creates, in pursuit of growth. In other words, the freedoms afforded by capitalism, like almost anything else in capitalist logic, come at a cost.

3.2.1 Cycles of Capitalism

To deal with the paradox of liberalism, capitalism has had to restructure and adjust itself on an ongoing, cyclical basis. To illustrate the dynamics of change, we follow a dialectical logic according to which the conflicts and tensions of a system drive change from one state to the next, giving rise to new conflicts and tensions that, in turn, become the drivers of change in the next state—a helical cycle that has been sustained so far, and will continue until the system exhausts its possibilities or until the tensions peak to a point of no return. Given the cyclic nature of the dynamics, we need to conceptualize them in the same fashion (figure 3.1):

1. Capitalism reinvents itself.
2. Socioeconomic predicaments emerge as a result of change.
3. Demands and crises arise due to predicaments.
4. Demands are co-opted, new opportunities are created, and the cycle repeats.

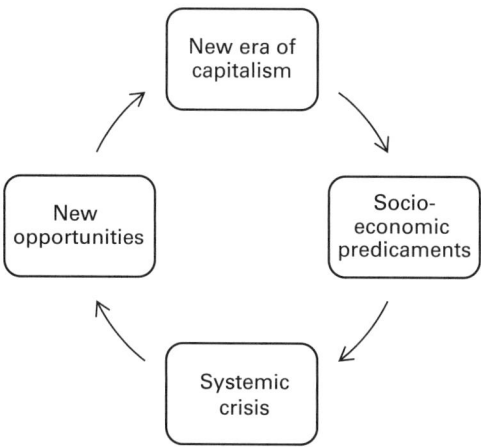

Figure 3.1

The dynamics of change in capitalism.

Source: authors.

Jameson (2011, p. 87) says, poetically: "Capital is an infernal machine, constantly breaking down, and repairing itself only by the laborious convulsions of expansion." Schiller (2014, p. 7) notes, "As regeneration takes hold, the seeds of subsequent crisis are planted deep in the political economy." Harvey (2010, p. 117) also observes the cycle, saying, "The crisis tendencies are not resolved but merely moved around." The "infernal machine" is hidden behind the normalization of neoliberal ideologies, but it is no more normal than slavery or feudalism.

3.2.2 Cycles of Individual Adjustment

The cycle of change at the systemic level has an equivalent at the individual level. Capitalism has mastered the art of control by devising new methods and refining old ones, expanding the scale and scope of control from direct supervision and bureaucratic mechanisms to indirect methods of quality control, market feedback, peer competition, and, more recently, ubiquitous mechanisms of surveillance and validation in digitized environments.

Consent, on the other hand, is broadly accomplished through what Antonio Gramsci (1971, p. 323) called "common sense"—the construction of a shared and taken-for-granted system of assumptions and beliefs out of long-standing social and cultural practices. One component of common sense might be the creation of the image of an "ideal subject" or "ideal worker"—an imagined person exemplifying the best worker/citizen within each era. In *The Protestant Ethic and the Spirit of Capitalism* (first published in 1904), Weber explained the notion of a capitalist "spirit" by invoking just such an individual: "If any object can be found to which this term [spirit] can be applied ... it can only be an historical individual, i.e., a complex of elements associated in historical reality which we unite into a conceptual whole from the standpoint of their cultural significance" (Weber 1904, p. 11). In *The Culture of the New Capitalism*, sociologist Richard Sennett (2007) similarly constructs the "ideal man or woman" who emerges according to the era.

The shifting nature of the ideal historical individual allows us to trace and make visible changes that happen imperceptibly around us (see figure 3.2). Such changes are more readily apparent when we step back to reflect on social, economic, and technical developments occurring in stages over periods of several decades. In the rest of this chapter, we outline these developments, with a focus on the distinct mechanisms of control and consent creation during each era, following changes at social and individual levels.

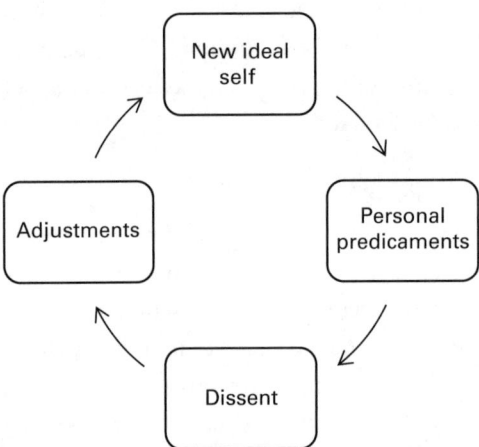

Figure 3.2

The cycle of individual change in capitalism. *Source:* authors.

3.3 Capitalism, Post–World War II–1970s: Massification, Satisfaction, and Dissent

Some time after the Second World War, observers of capitalism argued that the capitalist economy had given rise to a society where individuals were torn apart from their communities and put together in the shape of a flat, anonymous mass (Marcuse 1964). The war had disrupted communities, families, neighborhoods, marriages, romances, and friendships, creating a population now acclimated to disturbance and change. Capital leveraged this shift to take advantage of an increasingly mobile labor force, as well as new markets for ameliorative commodities that soothed the dislocations of being sundered from the familiar. The freedoms that followed independence from the watching eyes of old-fashioned family and community members eventuated in glamorous, exciting, and highly marketable goods and services.

3.3.1 Manufacturing Satisfaction

The technology of record during this period is, without doubt, the television (Fisher and Fisher 1996). In the late 1940s and throughout the 1950s, television became a fixture in the living rooms of virtually every demographic in the United States. Three channels provided entertainment for an entire nation. Such limited offerings seem bizarre today, but at the time, ABC, NBC, and CBS were wondrous sources of diversion, right in one's own home. Everyone watched the same (family friendly) shows, and mass entertainment created both national unity and a uniformity that would be fiercely attacked in coming decades. TV carried the chic and beauty of Hollywood into ordinary lives, stimulating aspirations for cosmetics, interior design, automobiles, and an intensive program of consumerist culture.

Disruptive shifts in personal circumstances, in particular separation from familiar others, were relieved to some degree through the satisfactions of televised entertainment, sports, and news. Television buffered the shuffling of people into new locales; neighbors getting to know one another in their new neighborhoods had something to talk about. In 1947, for example, the first World Series was broadcast on television (the Yankees beat the Dodgers), and the defining event of the "national pastime" brought together television's first mass audience, comprised of millions of viewers.

In 1956, Elvis Presley appeared on CBS's *Stage Show,* and a single performance rocketed the unknown rhythm and blues singer to overnight superstardom. "Heartbreak Hotel" topped the charts, and Elvis's extraordinary career as national icon and cult figure began. Perhaps only television could have brought the gritty, sublime Elvis to prominence, showing off the hip-swiveling sexuality, but, more subversively, delivering black-influenced music to mass white audiences in an era of considerable racism.

3.3.2   Voices of Dissent

Despite Elvis (an omen of cultural upheavals to come), most of the era's entertainment traded on wholesomeness and conventionality. Its conformist blandness grated on edgier members of the culture, in particular, folk musicians and the beat poets, who declared the satisfactions of mass culture insipid and dangerous. Mass entertainment and its consumerist residue induced passivity, sapping energy for critical thought in favor of trivial pleasures, said the poets. Songs such as "Little Boxes," written by Malvina Reynolds and popularized by Pete Seeger, critiqued the "little boxes all the same" orthodoxy of the middle classes. Allen Ginsberg, Gary Snyder, and Lawrence Ferlinghetti became famous for castigating an unimaginative, acquiescent bourgeoisie. Presciently, they lamented broken connections to nature. The poetry of Ginsberg epitomizes the movement, with the lines of "Howl" still echoing, their violence a counterpoint to the submissiveness they protest:

> I saw the best minds of my generation destroyed by madness ...
>   What sphinx of cement and aluminum bashed open their skulls and ate up their brains and imagination?

The riots of 1968 in Paris, which started as student protests and quickly expanded to general wildcat strikes, brought French society and (briefly) the government to a standstill (see figure 3.3a). The rioters' grievances are vividly captured in some of their mottos and graffiti:

> The more you consume, the less you live. Commodities are the opium of the people.

> Since 1936 I have fought for wage increases. My father before me fought for wage increases. Now I have a TV, a fridge, a Volkswagen. Yet my whole life has been a drag. Don't negotiate with the bosses. Abolish them.

We don't want to be the watchdogs or servants of capitalism.

Man is neither Rousseau's noble savage nor the Church's or La Rochefoucauld's depraved sinner. He is violent when oppressed, gentle when free.

We don't want a world where the guarantee of not dying of starvation brings the risk of dying of boredom.

In a society that has abolished every kind of adventure the only adventure that remains is to abolish the society.

These slogans illustrate the spirit of critiques made in the 1960s–1970s against the developments of previous decades—namely, desires for autonomy in the face of hierarchical control, liberation from boredom and lack of "adventure," and rejection of the idea of happiness as consumption of bland mass market goods. In the United States, similar critiques were expressed in the utopian streak of the hippie movement (Howard 1969). Alienated from the dominant culture, those who decamped to the Haight-Ashbury and Greenwich Village (see figure 3.3b) sought to provide an alternative vision of a society without the fetish of material acquisition.

By the end of the 1960s, folk songs and the beats' intellectual poetry had given way to the raw, angry populism of rock bands such as Country Joe and the Fish. Their signature "I-Feel-Like-I'm-Fixin'-to-Die Rag" was frequently heard on the radio, and was featured at concerts, including a famous rendition at Woodstock. The military-industrial complex (so

Figure 3.3

(a) Paris, 1968. Courtesy of Bruno Barbey. (b) San Francisco, 1967. *Source:* Associated Press.

named in the 1950s by President Eisenhower) came in for blistering critique as the technologies of rock music—ear-splitting amplification, electric guitars, stereos, and cheap vinyl—broadcast the message far and wide:

> Well come on Wall Street don't be slow,
> Why man this is war go go go,
> There's plenty good money to be made,
> By supplying the army with the tools of the trade,
> Just hope and pray that if they drop the bomb
> They drop it on the Vietcong.
> And it's 1,2,3 what are we fighting for?
> Don't ask me I don't give a damn,
> The next stop is Vietnam,
> And it's 5,6,7 open up the pearly gates,
> Well there ain't no time to wonder why,
> WHOOPEE we're all gonna die.[6]

If it sounds odd to identify speakers and electric guitars as crucial societal technologies, consider that at the time, there were no personal computers, cell phones, cable television, or handheld devices beyond transistor radios. Technologized music was the glue that united a generation, providing satisfaction for restless, uneasy masses, and salving some of the boredom and monotony the French students' harsh poetry conveyed. The music expressed—loudly and clearly—the fear, anger, and bewilderment of a generation for whom precarity meant out-of-control politicians and scientists of the bomb-dropping, body-box variety. Hiroshima and Nagasaki were very present to this generation, whose fathers and uncles had fought in World War II, and the threat of nuclear war seemed (and was) real. When Country Joe wrote his song, there was genuine fear of death. Indeed, another line from the Rag, addressed to parents, is:

> Be the first one on your block
> To have your boy come home in a box.

The image was not hyperbole; boys were coming home in boxes. Rock music, itself a child of capitalism, contributed nonetheless to a shared common space of protest and a place to work out frightening social predicaments. Caffentzis (2013, p. 93) notes the reappropriation of capitalist resources to fuel countercultural activities, observing that "[A]nticapitalist commons [are] founded on distinctly capitalist terrain."

While capitalism originates and embeds the activities of, for example, student protestors, hippies, and rock musicians, the activities may transcend capital's goals of control and accumulation of wealth. Caffentzis (2013, p. 92) remarks on the hoboes of the early twentieth century: "Through the complex organization of movement, information exchange, and reproduction nodes, the hoboes created a nationwide network that used the private property of the railroad companies as their commons." Hoboes appropriated, to a degree, what had been taken from them, refashioning capitalist resources for their own purposes.

Much as the French students had, hippies and other counterculture movements longed for liberation through exhilarating, authentic culture. The capitalist response to these longings was to recuperate them in intricate ways that allowed for increased profits and capital accumulation. Liberation was addressed through a strategy of stimulation that would satisfy people by mass commodification of their desires. The puritanism and constraints of the 1950s were met with new freedoms in the form of recreational drugs (illegal but profitable); sexual experimentation and the commodified sex of the pornography industry; and the revenue-generating pleasures of rock and roll. Loud, transgressive music was the touchstone of youth, and English and American rock bands and musicians (some of whom, such as the Rolling Stones, the Who, and Neil Young, perform yet today), ushered in an era that spawned the telling phrase "generation gap." Life got more interesting: exciting rock festivals happened; music lyrics spurned older generations ("Hope I die before I get old"), and ridiculed capitalist work culture ("Meet the new boss, same as the old boss"). Recreational drugs were plentiful and affordable, and established an ethos in which the coming use of legal psychotropic drugs in everyday life was normalized. Despite countercultural proscriptions against consumer culture, capitalism found ready markets for the selling of commodified desire, creating an economic boon that grew, and today includes lucrative industries serving international markets in film, books, television, magazines, video games, recorded music, and live concerts and festivals. When the Rolling Stones claimed they could get no satisfaction, their real point was that they were entitled to such satisfaction—a new development in a Euro-American culture previously characterized by relatively more restraint and modest expectations.

3.3.2 Massification

Capitalism continued to improve the economic well-being of a large segment of the population, especially compared to the worst failures of the previous era. In the United States, the outcomes of the New Deal and the implementation of a welfare state designed to ameliorate distortions of poorly regulated markets which were tangibly experienced by many in the Great Depression, were further enhanced in the 1960s by the policies of the Johnson administration toward equal employment opportunity. These accomplishments came by virtue of mass standardization of goods, services, and lifestyles, promoted vigorously by the new marketing and advertising industry. The freedoms of the new era were largely experienced in non-work contexts, and the constraints of bureaucratic control, which persisted in the firms and organizations of the postwar era, heightened sharp divisions between leisure, with its burgeoning satisfactions, and the sterility and regimentation of work.

The counterculture continued to resist this regimentation in its calls for tribal communitas evident in mass events such as rock concerts, and a shared culture of music in which everyone knew the latest hits and only a few genres of music filled every niche (unlike the splintering of music into too many genres to keep up with in the 1980s and 1990s). The combination of massification and satisfaction gave rise to a generational sense of empowerment expressed in various proposals for making the world a better place. This idealism depended on shared identity and a belief that collective actions could be intelligible to large swaths of a population, if not to everyone.

Because the slicing and dicing of markets into ever finer segments, and the hyperindividuality inherent in slogans such as "Army of One," had not yet occurred, a collective creativity responded to the serious problems capitalism was increasingly generating. It was during this turbulent era that sound proposals for dealing with disastrous environmental decline were put forward as programs of lifestyle change such as voluntary simplicity (Elgin and Mitchell 1977), the development of appropriate technologies (Schumacher 1973; foreword by Theodore Roszak), and the promotion of energy conservation. These proposals were driven out by the subsequent form of capitalism that emerged in the 1980s, but are being revived in the current era, in which climate change, pollution, species extinction, resource depletion, and other outcomes of intensive economic activity are increasingly troubling. In the 1970s, President Nixon initiated and/or signed into law the Environmental Protection Agency, the National Oceanic and Atmospheric Administration, and the Clean Air, Clean Water, and Endangered Species Acts. How much worse might our current environment be without this legislation? Some of the developments in the transformative 1960s and 1970s indicate at least some malleability of capitalist culture. Good ideas, whatever their immediate outcome, are cultural resources to be revived even after initial suppression, as is happening today (of which more in chapter 12).

3.4 Capitalism 1980–Present: Deregulation, Casualization, Surveillance, and Totalized Stimulation

In the postwar era we have just discussed, in "simpler times," so to speak, satisfaction and massification were tightly coupled, reinforcing and mutually producing one another. After the 1970s, deregulation of the economy, which broke the link between priorities for the common welfare set by the national government and economic activity, instigated an era of considerably more complexity. Several trends of interest to us—deregulation, casualization of labor, corporate surveillance, and totalized stimulation—occurred as multiple responses to capital's new freedom to do as it liked. Since 1913, the US government had seen to it that, for example, all Americans were served by telecommunications technology: "one system, one policy, universal service." Telephony diffused across all social divisions. Job protections were normative[7] as well as legal, and there were restraints on what could be shown on television, which was, of course, more easily policed, with its three channels, than the global internet of today. But all of that began to change.

3.4.1 Deregulation
Just as idealistic proposals for more regulation—to protect the environment, attain energy independence, promote rights for newly visible workers such as farm laborers—were gaining traction, capitalism seemed to snap out of a reverie, and energetically set about ensuring that restraints such as government regulation would not impede the project of capital accumulation. Capitalists aggressively asserted that the real problem was too much government. Increasing governmental efforts to improve the

collective welfare were reversed in the 1980s (de Brunhoff 2003). This reversal, which is at the foundation of neoliberalism, sought to organize, inform, and reform both the state and the society on the basis of market freedoms, and it brought about a number of shifts and consequences. Instead of the market and state mutually constraining each other, neoliberalism considered the role of the state as one that governs *for* the market rather than because of or despite the market—that does not, in other words, interfere with the effects of the free market. In brief, neoliberalism dictated a social policy, the purpose of which was not to nullify the potentially destructive effects of a market economy on social life or the environment, but to ensure competition and to formalize society on the model of the enterprise (Foucault 2008). In this fashion, not only did every individual now embody an enterprise—indeed each individual was urged to consider herself a "brand"—the enterprise itself was transformed from the family-owned, hierarchical firms of the previous era to an institution run by a new generation of professional and academically trained managers with MBA degrees.

On the basis of these shifts, the social policy prescribed by neoliberals was one that eschews security and equality as its objectives, and suggests that any resulting precarity is the fault of the individual who does not work hard enough, or choose the right education and career path, or forgo having children until the moment at which all is in order. In dealing with risks such as illness, accidents, and natural disasters, this policy leaves individuals to their own devices and sees to it that most—though not all—have enough income to insure and protect themselves, at least minimally (Beck 1992). Those who fall below the bar have only themselves to blame, according to this perspective.

The key policy mechanisms of the neoliberal agenda are privatization, deregulation, and withdrawal of the state from many areas of social and economic provision (Suarez-Villa 2012). The ideal person who would emerge in this society is not a consumer, but an entrepreneur with her own brand who depends on stock markets and personal productivity for security—a realist individual cognizant of the value of competition and independence (ibid.).

An early bellwether of change in risk allotment was the passage of the 401(k) law in 1978. In the beginning, 401(k)s were a way for employees to defer tax on the portion of income received as deferred compensation, so that corporate executives could supplement their income. Corporations such as Hughes Aircraft, Johnson & Johnson, PepsiCo, JC Penney, and others immediately saw the cost-saving handwriting on the wall and began making plans to shift worker retirement accounts to 401(k)s, though this had not been the instrument's original purpose. Congress approved worker plans in 1981. The number of plans with a 401(k) feature continued to grow, steadily replacing traditional pensions, and the amount invested in 401(k)s today is in the trillions of dollars.[8]

3.4.2 Casualization of Work

Free of many of the curtailments of regulation, capital accumulation grew through the casualization of work. (By "casualization," we mean that the employment relation relieved employers of long term commitments to see workers through to the end of their careers and provide benefits such as insurance and paid vacation.) Capitalism was experiencing its own moment of liberation, cutting loose from expensive societal expectations for worker protections that had prevailed for decades, such as job security and pensions (de Brunhoff 2003). In firms and organizations, units and divisions were replaced with project teams given autonomy through the creation of profit centers, quality circles, and other new organizational forms (Caffentzis 2013). These forms sought to transform employees into entrepreneurs, whose job tenure was now a function of their individual performance under exacting monitoring. One effect of these changes was an increase in the number of "casual workers"—temporary workers, part-time workers, contract workers, the underemployed, the laid-off, the self-employed, legions of freelancers and consultants—who made up a flexibly available workforce (de Brunhoff 2005). The efficiencies of slotting the right worker into the right task at the right moment—in a context in which rapid response to dynamic markets was key—is not to be underestimated.

In part, this shift occurred because commodity markets were changing: durable goods such as refrigerators and automobiles, purchased only occasionally, had saturated the market and were now joined by an array of commodities whose fashions shifted rapidly—from designer glasses to upscale restaurant food to electronics (Gordon 2012). The easy availability of credit and the relaxation of norms of frugality fueled spending in the new economy, provoking a heady sense of freedom even as it pushed people to precarious fiscal decisions that might eventuate in something more like indentured servitude than freedom, as those in debt struggled to cope with high-interest loans that never seemed to be paid off (de Brunhoff 2003).

In this era, mobility became a key value and criterion of evaluation, and loyalty (to the job, the firm, the employer) was discounted as a sign of incompetence and lack of skills (Ekbia 2016). New work arrangements brought about new control mechanisms that utilized team members to monitor each other's performance. Increasingly computerized, these mechanisms created work environments with a potential of intensive control that extended beyond the immediate working hours. A combination of self-control, peer monitoring, and computerized surveillance gave rise to a novel form of control that is less visible but also more continuous and more intense compared to previous eras.

3.4.3 Surveillance

Unlike the televisions and record players of the previous era that gave up no secrets about their users, a culture of digital scrutiny began to emerge that has, since the mid-1980s, contributed to anxieties about the integrity of our personal information, and, indeed, our lives and identities. This change was manifest in a small industrial project that suggested how useful such scrutiny could be. Active Badges, a location-monitoring service, instigated a privacy debate that continues today. Developed by researchers at Xerox PARC, Olivetti, Metaphor Computer, and Sun Microsystems (key players in tech at the time), Active Badges tracked workers' locations and reported them to a centralized system. The abstract to the first paper about Active Badges read: "A novel system for the location of people in an office environment is described. Members of staff wear badges that transmit signals providing information about their location to a centralized location service, through a network of sensors" (Want et al. 1992).

Behind the bland prose, the shocking implications were clear, and caused heated discussions framed in terms of privacy, security, surveillance, and technical control over data about oneself. Humorous commentary of resistance, such as "We don't have to show no stinkin' badges," was soon replaced by less humorous commentary at Olivetti and EuroPARC, where the badges had been deployed, when employees decided they did not want to wear Active Badges after all. Roy Want, who helped develop the technology at Xerox PARC, recalled, "Later the trend reversed, and equally quickly people did not want to wear badges anymore" (2001).

The technology broke taboos against centralized collection of personal location information—information people did not necessarily want visible. A mere 10 years later, Scott McNealy, CEO of Sun Microsystems, famously said, "You have zero privacy anyway. Get over it." While not everyone agreed, the fact that it had become possible for the CEO of a leading technology company to make such a totalitarian statement foretold the dawn of a new era. It is, after all, exposure of any aspect of one's life to the scrutiny of powerful others that upholds totalitarian regimes. Governments cultivate organizations such as the Gestapo, Stasi, KGB, Tonton Macoutes, and Savak to control the populace. Recent activities by the NSA leave many furious. The tradeoffs evident in Active Badges between utility (for example, locating employees efficiently), and the troubles pursuant to becoming visible to unseen others, foreshadowed current social quandaries.

A former employee at an AI startup, IntelliCorp, in Menlo Park, California, recalls how quickly things changed and what it was like to go from limited networking to the internet in a few short years:

> In 1984, the entire company worked through one 9600 baud modem to Stanford. 1984 was early for TCP/IP, but I know that I was writing documents using EMACS over a live link to a SUMEX computer and so it must have been TCP/IP. It wasn't ARPAnet, though. I remember it was a big deal when [one of the managers] got the ARPAnet link going. It was a big deal because we were prohibited from speaking about purely commercial topics over ARPAnet, so I had to throw random DoD [Department of Defense] questions in to be able to exchange email with Sperry, questions like "I hope your DoD-related users in San Diego aren't having problems with KEE [IntelliCorp's product]. Let me know if they do." When I neglected to do so, [my manager] cut me off from email until I promised to do better ... And then, a year or two after that, ARPAnet got changed to Internet and suddenly it was OK to talk about commercial topics in email. (Doug Shaker, email, personal communication, 2012)

As Shaker's story indicates, networking rendered workplace surveillance efficient; managers could easily monitor email (and other documents) and assess the adequacy of employee performance in nuanced ways. The digital machinery for surveillance simply appeared on our doorsteps. No one mandated, designed, or planned it; it arrived, "for free" as it were, as a powerful, if unlooked for, capacity of networked technology. Big Brother gained ascendance more in the manner of the market's invisible hand than as the calculated force of Orwell's *1984*. Without even noticing, we were taken up into networks in which, inescapably, we became visible to others. At work, we became accustomed to such visibility.

Not all efforts to surveil were as simple as reading email. Boltanski and Chiapello (2005, p. 248) observed that once managers recognized the capacities of computerization, they equipped themselves with "much more sensitive instruments of control," capable of measuring performance at the level of the firm, the plant, the team, and the individual. Such measures, often built into Enterprise Resource Planning (workflow) systems, were many times more powerful as a result of being networked. Previously, performance data were housed in "disconnected databases," making data integration across sources cumbersome. Networked databases afforded increased oversight in occupations such as sales, where detailed information about customers and products could be collected, allowing managers to fine-tune performance evaluations with information made visible through technology (Boltanski and Chiapello 2005).

While such surveillance unsettled people, on the flip side, the pleasures of networked technology were immense: this was the era of augmentation in computing we discussed in chapter 1. Capitalism was both giving and taking, as usual.

3.4.4 Totalized Stimulation

In the previous section, we discussed the buildup of products aimed at commodified desire. Today, we seem to have reached a fever pitch of such products, which are available 24/7. (A random read of Google News nearly always turns up stories involving commodified drugs and sex[9].) Escaping monotony is a compulsion—or maybe a pervasive cultural pathology to which we have become habituated.

More important, such escape is a consistent cultural response to the demands of work. The level of autonomy and engagement with work in the capitalism that began in the 1980s requires a kind of intense subjectivity that is fully immersed in activity, drawing the whole individual into the stimulating process of high productivity and performance. The new era needed information about individual behaviors, tastes, and preferences, among other things, in order to keep individuals involved and engaged. Largely enabled by the inexpensive processing power of modern computing, the new arrangements actively sought this information directly from individuals themselves and indirectly by surveilling their behaviors and actions. A huge expansion of information voluntarily and involuntarily provided by people—a kind of "information through communication," as opposed to a panopticon's information in the absence of communication (Foucault 1977)—suffused work life, and then home life.

Today, we are familiar with the vast quantities of data collected about us as we instigate personal programs of totalized stimulation: participating in social media, playing games, watching movies, reading the news, searching for information, contributing to forums, posting pictures, shopping, and so much more. We are aware, sometimes dimly, sometimes more self-consciously, of how much time we spend online. A tagline from a Yelp contributor charms with its self-deprecating humor: "It's Friday night ... Why am I home alone Yelping?!?" The poster notes his aloneness without self-pity, indicating engagement with his Yelp tasks. His stance toward separation from others is ironic, distanced, ambivalent. Though it is always difficult to understand one's own era, we are in the midst of change so rapid it creates stupefying distortions (Virilio 1986). The Yelp poster's wry commentary and the layers of meaning beneath it are familiar to all of us who spend significant time online.

3.5 Summary

We have used this brief, impressionistic history of two eras of capitalism to foreground what is usually left out of such accounts—the everyday technologies of home and work, and how they draw us into their orbit. Through complex, technologically mediated forms of consent and control (such as totalized stimulation and surveillance), we enter networks and remain there, providing labor. In the next chapter, we look in more detail at exactly what happens as we participate in these networks.

4
Possibilities and Predicaments: A Story of Stimulus

4.1 Introduction

In this chapter, we examine processes through which the masses come to participate in computer-mediated networks. The *possibilities* within these networks are well known to all of us, and underpin capitalism's techno-utopian dreams of economic prosperity, universal education, political participation, self-expression, and shared ideas and projects. Aspirations for active engagement with technologies have provided ample opportunity for the invention and introduction of an array of new products and services. We are more informed, connected, educated, empowered, and entertained than ever before—thanks, in significant measure, to digital technologies. This is the positive and productive side of engaging with computers.

There is, however, a dark and negative side to the story as well. The *predicaments*—separation, precarity, futility, and monotony—incite participation in networks as we attempt to find some relief from and practical remediation for the struggles of everyday life. In characterizing the predicaments, we have drawn broadly from the work of Gorz (1985), Giddens (1990), Beck (1992), Rifkin (2000), Hornborg (2001), Holmes (2002), Boltanski and Chiapello (2005), Sennett (2007), Wright (2009), Harvey (2010), Caffentzis (2013), Schiller (2014), and others who describe conditions of contemporary capitalism. We draw also from our own empirical studies of video gaming, healthcare applications, and social media.

We use the predicaments to explain processes of heteromated labor that do not, on their face, make sense. There is not an obvious logic for performing labor that has no expectation of reasonable (or any) remuneration. Capitalism typically supports necessary labor-time so that workers can sustain themselves and participate in the economy. Heteromation is something of an enigma in this regard: it generates economic value for enterprises, but seems to defy expectations of both necessary labor-time (compensation sufficient for subsistence) and labor power (the commodity laborers sell to capitalists). Instead of selling labor power to sustain ourselves and to act in the market as customers, we provide labor for free, or at such a low price point that it cannot be commensurate with necessary labor-time.

What, then, drives us to perform the heteromated labor that contributes to the immense wealth being amassed by Google, Facebook, Amazon.com, video gaming companies, healthcare enterprises, and a host of other corporations? We attempt to answer this question through consideration of the anxieties and agitations the predicaments impel. We examine the ways in which digital technologies may calm or ameliorate the struggles. In doing so, we avoid modernist tendencies to psychologize—that is, we do not assess outcomes of the predicaments as resulting from particular, individualized personal events or personal histories, but rather as shared material experiences of life under contemporary capitalism. Predicaments, as we define them, are not psychological in nature, and they cannot respond to psychological interventions such as therapy. The struggles do respond to culturally shared resources, in this case, digital technologies. The success of computer-mediated interventions varies, sometimes—perhaps often—producing satisfaction, and at other times failing to deal with what ails us, or perhaps even making it worse.

4.2 The First Predicament: Separation

Capitalism pushes us to be mobile, keeping us on the move as we recover from job losses, pursue education, and follow spouses and family members doing the same. Sometimes we move to seek our own bliss, responding to visions of the heroic individual promoted by the neoliberal ideology. Separation from friends, family, neighbors, and acquaintances is the inevitable result of mobility. The varied forms of separation come with many costs: loss of community, loss of connection, and sometimes loss of identity. Separation from loved ones formerly within range of face-to-face contact, reduction in opportunities for participation in geographic communities, and loneliness and boredom are common phenomena of modernity. Stories of the trials and tribulations of immigrant families dealing with separation are chronicled elsewhere, but they are not unique to those recognized as immigrants. Similar phenomena can be found in situations when the individual is not prepared for a move, as happened with a graduate student who wrote the following to one of the authors:

> I've been having an extremely hard time this semester with my anxiety and depression and being away from everyone I know. It's the last week of classes and everything is falling apart. I'm completely overwhelmed and scared. I've been hiding in my room ... Being alone, without any support system, in a large school has been too much to handle.

Loneliness has been a recurring theme in American literary imagination. Robert Ferguson argues in *Alone in America* that loneliness has a special place in the United States because of the "the openness, mobility, uncertainty, and flux in (what was originally) a spacious new country" (2013, p. 2). Following the theme from the perspective of "dissolved domesticity" in the works of authors such as Washington Irving, Nathaniel Hawthorne, Louisa May Alcott, Mark Twain, and Henry James, Fergusson presents the different faces of loneliness as "failure, betrayal, change, defeat, breakdown, fear, difference, age, and loss."

Loneliness also has other meanings having to do with loss of identity. Identity concerns one's perception of one's place in the world. A key component of this perception for modern persons comes from their careers—a key mechanism for constructing and maintaining identity is to be able to create a meaningful narrative of one's life, including work. Stable lives, activities, connections, and communities are the basic ingredients of such narratives. Their disappearance generates a strong sense of loss of identity (Sennett 2012). The consequences of separation may thus cut to the core of the contemporary subject, undermining the very notion of who a person is, causing feelings of "falling apart," as the student wrote.

The loss of a coherent narrative about oneself, as well as the practical realities of being "without a support system," produce voids into which digital technologies rush. We use them to patch up our "dividualized" selves, as Deleuze called modern persons. Social media and multiplayer games spring to mind in this context; they allow us to connect to and communicate with distant others. A darker side is evident in the peculiar disinhibitions the internet spawns (Kiesler, Siegel, and McGuire 1984;

Sproull and Keisler 1986; Suler 2004; Sood, Antin, and Churchill 2012), manifest in, for example, trolls and malicious hackers whose actions seem rooted in an anomie of loneliness and separation. The homophobia, misogyny, and racism gaming communities put up with, even in the midst of the restorative and validating activities that make the games worthwhile, continue to be problematic (Nardi 2010; Huntemann 2014). These expressions of hate flow, in part, from the anger of the dividualized self.

4.3 The Second Predicament: Precarity

The human condition has long been characterized by "problematic situations," as Dewey wrote long ago (1931). These situations have, however, acquired a novel scope and dimension in contemporary life, largely deriving from the gradual disappearance of institutional support structures and the undermining of the welfare state, leaving the individual in a condition of economic and professional precarity.

Recent financial crises of 2001 and 2008 provide ample evidence of fragility and uncertainty. Policies adopted by banks, investment firms, auditing organizations, and other economic entities create ongoing risk in financial markets. The economic models and practices of these organizations implement and propagate risky decision-making processes that destabilize the economy (see Snider 2014). Although notorious cases such as Enron in 2001 and Bear-Stearns in 2008 received a great deal of press, risk-taking behavior has been shown to be endemic to the functioning of today's capitalism (Beck 1992; Neff 2012; Suarez-Villa 2012). Astonishingly, this type of behavior does not stop at the level of the corporation, but moves the unit of "risk management" to the household and individual, where people are literally pulled into perilous scenarios and situations, the appreciation of which is beyond their knowledge and understanding. Risk is pushed down to individuals who must manage complex decisions with respect to taking loans, planning investments, managing healthcare, creating a strategy for retirement, making decisions about how to handle eventual disability or chronic illness, and much more.

Precarity of employment resulting from capital's need to respond quickly to fluctuating market conditions (and sometimes its own poor planning) creates a workforce of millions of underemployed, unemployed, self-employed, and part-time workers (Standing 2011). They manage risk by constantly retraining, rebranding, and reconfiguring their careers. The notion of "trickle-down economics," advocated by neoconservative pundits and politicians, makes much more sense if it is understood as pushing risk, rather than benefits, down to the lower strata of society. An obvious example was the housing bubble of 2008, when people were enticed by high-risk, variable-interest loan structures to buy homes and properties significantly above their means.

Other critical aspects of contemporary life, such as pensions, education, and healthcare, have fallen into the same dynamic, forcing individuals into the adoption of risky behavior, or, in the more innocuous scenarios, forging them into units of risk management. The implementation of the Affordable Care Act is an instance of the latter, where individuals are legally required to find the most suitable health insurance plan from a range of options in a marketplace of health exchanges (if they do not get health insurance another way). While the dominant discourse is shaped around giving more people access to healthcare, the law can also be understood as an attempt to detach healthcare from employment, as employers seek to reduce bonds of commitment between themselves and employees, and to continue to push the development of the independent, "free" neoliberal person, responsible for planning and managing his or her own life.

Funding and planning that used to come from government and corporate sources are diminishing, forcing us to figure out how to go forth and manage these needs ourselves. Increasing burdens of self-management create confusion (which health/retirement plan should I choose?), time pressure (the deadline is soon), fear (what if I choose wrong?) and anxiety (how will I pay off my student loans?). Living under a shadow of precarity and unpredictability, we undertake activities we hope will be a bulwark against disaster by establishing a personal brand, educating ourselves, retraining through programs of continuing education, trying to stay healthy (in the face of constant temptations the economy sets in front of us), creating and maintaining a reputation to ensure professional opportunities, self-funding retirement, and so on. Digital technology generates rich sources of information about these activities and decisions, which are ongoing and mutable, in need of constant update and maintenance.

4.4 The Third Predicament: Futility

Another outcome of the uncertainty of employment and the unstable life narratives that separation and precarity contribute to is a pervasive sense of uselessness and futility. This sense is reinforced by the inability of educational credentials to provide persistent skills and employment opportunities in the job market. Some people experience permanent unemployability due to age, disability, or the need to provide care to aged or infirm family members.

Career disappointments may be acutely traumatic.[1] Too many workers find that lack of opportunity obviates escape from a lifetime of dead-end jobs and economic insecurity, or that advanced education fails to deliver the hoped-for result. The increasing use of perilous payday loans, purchase of lottery tickets, and the abandon with which people acquire luxuries such as costly engagement rings, massive TV screens, and expensive smartphone services, indicate that we may slip into feeling that it does not matter what we do. At the peer-to-peer lending site Prosper.com, for example, the investor chooses among several loan types, the most common of which is "debt consolidation." The desperation driving people to seek loans to gain control of metastasizing debt, and the determination brought to these grim projects of fiscal discipline, stir our sympathy. Remarkably, other loan categories include "engagement ring financing," "vacation," "boat," and "household expenses." A vacation loan at the borrower rate of 13.85 percent and a monthly engagement ring payment of $460.20 are typical examples drawn from Prosper.com (the name of which sounds increasingly ironic the more one scans the loan applications).[2] Capitalism urges us to consume beyond our means, and then stands by ready to pick us up off the floor when we need to "consolidate debt," or we have been persuaded to consume even more things we don't need through easy credit. As Boltanski and Chiapello (2005, p. 435) say, capitalism "takes back on one level what it offers on the other."

4.5 The Fourth Predicament: Monotony

The human experience of monotony probably dates from the time when we learned to use tools and talk to one another. So stimulating are these uniquely human capacities that we have likely acquired a hardwired taste for physical and cognitive stimulation unmatched in other animals. Today, our culture and economy are geared toward intense cultivation and exploitation of our penchant for stimulation. Engagement with products to satiate commodified desire drives people to continually seek an array of comforts or excitements offered online. The high quality and easy availability of entertainment, beginning in the 1950s with the mass production and distribution of televisions and record players, continued in subsequent decades, culminating in the present time in which a profusion of music, film, TV, games, commentaries, opinions, news, gossip, and much more, is available online, on demand. Rather than a pleasant accessory to life, such diversions may become cravings that urge us to continually seek intense, constant stimuli. In 2013, Timothy Wilson, a psychologist at the University of Virginia, reported that many subjects in his experiment chose to administer a painful electric shock to themselves rather than sit quietly, with no external stimulation.[3]

The need for constant stimulation encourages people to treat basic aspects of personal and social life as transient and fleeting encounters that need to be dealt with as they arise: marriage is a short-term contract rather than a long-term commitment; the rituals and restraints of romance are superseded by hooking up and pornography; being informed is checking the headlines of customized news streams and tweets; reading is squeezed into short bursts, elevating graphic novels and brief online pieces or relying on data mining technologies that can give the "gist" of a book without the hassle of a close reading; thinking is a luxury of the past that can only be afforded by armchair philosophers of yesteryear, or an experience too drastic to endure, even if electric shock is the only alternative. Jonathan Schooler, a professor from the University of California, Santa Barbara, commented on the University of Virginia experiment: "It seems that the average person doesn't seem to be capable of generating a sufficiently interesting train of thought to prevent them from being miserable with themselves"(Francis 2014).

If we contextualize the experimental results within the predicaments, it is less surprising that a desire for stimulation—distraction, diversion, sensation—accompanies experiences of separation, precarity, and futility. We would rather not think about them! Amplifying the effect is the well-known phenomenon of habituation, which requires increasingly higher levels of a stimulus to reach the desired state. The constant stimulation of endless content on the internet is a powerful antidote to monotony, yet drives a compulsion for even more stimulation. Few of us with internet access do not turn to its copious offerings for distraction, diversion, and sensation. For the first time in human experience, highly stimulating materials are available to all, any time of day or night, any day of the year. We do not yet know the long-term implications of the availability of such stimulating materials, but it is undeniable that we seek them out.[4]

4.6 Summary

After considering the predicaments, we see that the "light" and "dark" sides of neoliberal capitalism are not simplistically black and white. We shift between them at different temporal scales, some momentary, some more enduring. For example, Bonnie's studies of video gaming demonstrated a plethora of positive social, cognitive, and psychological outcomes, yet gamers themselves were the first to ask about the obsessive behaviors they found themselves and others exhibiting, some of which they were actively attempting to forestall in their lives, or in the lives of others (see Nardi 2010, chapter 6). Anyone who has played a beloved video game knows that the border between passion and mania is not an easy one to defend. Dewey (2005) describes how our passions may "overwhelm" us.

Hamid's studies of health technologies such as personal health records similarly illustrate their ambivalent character. On the one hand, with their potential for tracking and communication of health-related activities, these technologies can empower patients, allowing them to be informed participants in taking care of their health and in their relationships with doctors. On the other hand, they add the burden of data management onto people who are already overwhelmed by their health conditions, offloading new forms of work onto individuals. The objectification of individuals that derives from these interventions epitomizes the kinds of predicaments that efficiency-driven approaches can generate (Ekbia and Nardi 2012).

Our goal in enumerating the predicaments has been to problematize easy techno-utopian conceptions of the simple utility and beneficence of technology, and uncritical celebration of "innovation." A broader picture must insert social and economic conditions such as unemployment, learned boredom, outdated credentials, and so on, into the processes that fuel technological participation.

We turn now to presentation of specific cases to see how heteromation plays out in communicative, cognitive, emotional, and organizing labor. Each of these types of labor draws on particular human capacities, inserting living labor into billions of moments of heteromated value creation.

II
Varieties of Heteromated Labor

5
Communicative Labor: A Story of Separation

5.1 Introduction

The human capacity to communicate is at the core of our sociality. As a species, we are social to the extent that we communicate (Levinson and Enfield 2006). This truism has acquired a special meaning in the age of the internet, when a great deal of our communication is mediated by computing platforms such as email, social media, electronic commerce sites, and all the rest. The historical novelty is in the integration of multiple forms of communication into the composite and interactive medium of the internet. The bigger novelty, however, is in the fact that a large portion of this medium is now owned and operated by private corporations that have emerged as global powers in the last couple decades. A few numbers can put this into perspective.

From December 2008 to December 2013, the number of users on Facebook went from 140 million to more than one billion. During this same period, Facebook's revenue rose by about 1,300 percent, with ad revenue rising from $300 million to above $4 billion. Facebook's filing with the Securities and Exchange Commission in 2012 indicated that "the increase in ads delivered was driven primarily by user growth" (Facebook 2012, p. 50). LinkedIn, similarly, has enjoyed exponential growth (300+ million members as of February 2015), but most of its revenue comes from "a small group of recruiters looking for talent," along with advertising income and fees for its Premium Membership[1]. Twitter, with an income of $1.3 billion in 2014, has banked on our desire for validation: "What you say on Twitter might be seen around the world instantly." Schiller notes that in 2013, Google "announced to users of its YouTube, Google Play, and other services that their photos, profiles, comments, and rankings might be deployed in constructing advertising endorsements across the two million Web sites served by its display-advertising network, sites that are viewed by an estimated one billion people" (2014, p. 173). Validation comes at the price of corporate ownership of data we choose to share, such as photos and personal news, but also extends to the manufacture of data we have not shared, i.e., data constructed about us, such as rankings and classifications that use the raw material of our posts and clicks to tell stories about us that are not of our own making.

Many startups have also emerged in the last few years with business models that monetize online analytics—i.e., the data extracted from patterns of communication on social media and other platforms. Pulling together data from Facebook, LinkedIn, Twitter, and other sites, a service such as Klout, for instance, provides a score between 1 and 100 for an individual's "influence" in terms of the size of their network, its content, and the number of users interacting with the content. Another company, Acxiom, describes one of its lines of business as global media companies "who grow revenue and market with *easy-to-sell, high-value audiences,* using the best possible consumer data and accurate, permissible data matching services in a privacy compliant manner" [emphasis added].[2] Providing information to corporations, nonprofits, political organizations, and government agencies, Acxiom reassures us—citizens, consumers, donors—that we "get more relevant advertising and special promotions and [that this] reduces the chance that you'll become an identity fraud victim." A few lines below this assurance it is mentioned that this information might, however, include "sensitive information such as the Social Security number."[3]

The direct relationship between the growing number of users and the expansion of these large and small companies is a well-accepted verity among observers—indeed, some seem to plainly state that heteromated labor extracts significant value (without, of course, using the term). According to the VP for Research of the technology consulting firm Gartner, Inc., for instance, "Facebook's nearly one billion users have become *the largest unpaid workforce in history*" (Laney 2012, emphasis added). Writers who analyze this workforce from a Marxist perspective consider the mechanism of value extraction to be the same as the traditional capitalist exploitation of surplus value. Fuchs (2011), for example, argues that users of social media are infinitely exploited because they contribute their labor-time without any reimbursement. This argument is based on the earlier theory of "audience commodity," according to which users of media (e.g., TV watchers) are sold as a commodity to advertising companies and their clients (Smythe 1981). Other scholars find the mechanism in the capability of companies to aggregate the "affective investments" of users (Arvidsson and Colleoni 2012), and still others the commodification of personal data (Andrejevic 2015; see Profitt, Ekbia, and McDowell 2015 for a compendium of these views).

5.2 The Networked World: Mobility, Separation, and the Desire to Connect

We consider the mechanism of such value extraction as a form of heteromation that allows both securing value and making it invisible through data production. The interposal of privately owned computer networks into human communication has turned interpersonal communication into a kind of labor, the economic benefits of which go to corporations. In the "networked world," people are valued to the extent that they can forge new links, and, to some extent, remain distrustful of (and, hence, disloyal to) pre-established structures and institutions. This is a world where forging relationships can be a source of benefit ("social capital"), and where people's value derives in part from their degree of mobility. Mobility, however, comes at a cost—most obviously, of losing one's footing in community, with the attendant predicaments of separation and status anxiety we discussed in chapter 4. These predicaments, reinforced by the uncertainties of unstable working environments, provide a powerful incentive for staying in touch and the underlying push toward social media.

Web 2.0 technologies respond to this situation by providing a space of interactivity that purports to fill the gap of separation. The push of predicaments is complemented by the pull of technology, which pragmatically and programmatically enables remote and persistent connection. In this fashion, interaction in the networked world brings a diverse set of actors—manufacturers of networking technologies, software development houses, ISPs, and social networking sites, along with "users"—into contact with each other as nodes in the same network. In so doing, it turns communication into a computable activity that produces rewards for the individual and economic value for the other actors.

5.3   Social Data: Value Extraction through Communicative Labor

Web 2.0 computing platforms are, by and large, dedicated to the production of *social data*—that is, content generated by users through online interaction and communication. Interacting with others in these environments through "tagging," "sharing," "liking," "upvoting," and so on, involves actions that produce, as a by-product, different types of social data: descriptive (e.g., individual profile data), behavioral (e.g., "liking"), and user-generated (e.g., photos, videos, comments). Social data is different from other kinds of data in its character and in how it is created, and both of these differences are important for understanding its economic value.

5.3.1   What Is Social Data?

In terms of its content, social data is an innovative concept. Unlike sociodemographic data that have been traditionally collected through the institutionally sanctioned expertise of statisticians, demographers, accountants, marketers, and epidemiologists, social data is collected from activity in online environments. Rather than being *about* social behavior, in other words, it is produced *through* social behavior, acquiring a fine granularity that was absent from the aggregate statistical data of yesteryear. Facebook's EdgeRank algorithm, for instance, analyzes around 100,000 parameters in order to push the right type of content (what to watch, whom to add, and so on) to a user. Social data is, therefore, also personal in at least two senses—first, in the way it is organized, presented, and "personalized" (van Dijk 2013), and then in the sense that its effects are fed back to the individual through algorithmic analysis and data mining (Andrejevic 2015).

The transition of personal profiles on Facebook from collections of data items to the narrative structure of a Timeline highlights the first point. Through a retroactive chronological construction of individual life stories and milestones (births, graduations, weddings, and so on), the Timeline provides an organized, digital version of the venerable "shoebox," from which selected bits and pieces of one's life are compiled and displayed for the world to see. At the same time, the transition has enabled new techniques of pushing data to users—e.g., through PageInsight data that allows marketers to obtain real-time analytics about user behaviors and preferences. As van Dijk (2013) argues, online platforms' advocacy of a unique online identity for users derives from their vested economic interest in this aspect of personalization.[4]

This business concern also relates to the other personal aspect of social data—namely, its capacity to be fed back to the individual in the form of ads and recommendations, as well as the distribution of opportunities such as employment.[5] This capacity is behind the idea of "data as asset," famously announced at the World Economic Forum in 2012 as a maxim of contemporary capitalism. This idea underlies the business model of companies such as Acxiom, which, with major clients such as United Airlines, Macy's, and Epson, boasts data products such as "Audience Propensities":

> While other big data analytics focus on what was, Audience Propensities focuses on what is likely to be. The result is more effective marketing, higher profitability and a superior customer experience.[6]

The statement speaks truth to a historical fact about the need of capitalism to speed up the circulation time of products to increase profits. Marx noted a long time ago: "The main means of cutting circulation time has been improved communications" (Marx 1990, vol. 3, 164). The mediating role of commercial media is, therefore, not new from the perspective of market efficiency (Comor 2015), but what is new is the capability to forecast "what is likely to be," as limited and flawed as this might turn out to be (Ekbia, Nardi, and Šabanović 2015). As Burdon and Andrejevic (in press) note, users are often unaware of the full scope of the uses of their data. They are unaware that they have consented to its collection and usage. For instance, Burdon and Andrejevic studied technically savvy users, finding disconnects between corporate data collection practices and people's perceptions of those practices. The authors noticed, for instance, a tendency on the part of their study participants to consider the high volume of data collected as providing anonymity—a tendency reinforced by statements such as Google's: "no human reads your email" (Webb 2004). Burdon and Andrejevic observed that:

> While it is true that big data firms such as Google may not be interested in particular individuals per se, it is certainly interested in collecting their data in

order to aggregate it with information from other users as a means of more effectively managing and manipulating them—at the individual level. In other words, both of the following claims can be true simultaneously: (a) that Google is not interested in particular individuals but that (b) it will nonetheless collect and store detailed information about users in order to more effectively tailor advertising and other forms of content to them, and indeed to the general population of users. In that sense, lack of interest does not mean lack of impact.

Social data, while maintaining an innocuously impersonal appearance, acquires a very personal character at the same time.

5.3.2 How Is Social Data Collected?

Social data is also different in *how* it is obtained. What might seem to be "natural" forms of communication are, in fact, carefully tailored modes of interaction intended to generate data footprints. By programming user participation so as to make interactions computable, Web 2.0 platforms collect data in ways that are distinct from the generation of data through automated technologies of monitoring, data tracking, and recording. This presents an interesting instance of how heteromation trumps automation, surpassing it when possible and subsuming it when needed.

The standardization of the interface is the first step in this direction. By providing specific placeholders (e.g., for photos of particular life events), for instance, the Facebook Timeline cues the user as to the type of content they should include on their profile. In this manner, these platforms create and promote a special kind of user—a user who "shares" (John 2012; Shilton 2016).

The standard interaction templates go further, formalizing everything on the interface (users, photos, posts, comments) as "objects" related to each other through pre-established actions such as "following," "sharing," and "clicking." As Alaimo and Kallinikos argue, "behavior-related data is thereby obtained by encoding social interaction *qua* action into a data-connection between objects." Rather than simply "capturing" data, in other words, "social media encodes, formalizes, and constantly manipulates data" to produce a new kind of sociality. What makes these encodings novel is that data, rather than top-down categories based on established taxonomies of education, income, demography, and so on (e.g., a male applicant with a college degree and annual income of $60,000), dynamically put people in transient and invisible buckets such as "a good borrower," "a bad insurance payer," "a promising job applicant," "an avid reader of mystery novels." The translation of these dynamic categories into specific business practices of economic value, such as employee recruitment, product placement, product development, and market strategy development, provides value to corporations that buy the data from social media companies and other operators of Web 2.0 platforms. At the same time, however, the complexity of the analytic processes that generate the categories makes them opaque to the user.

5.4 The Invisible Value of Social Data

The participation of the masses in networks where communicative labor takes place presents an ideal person as someone who is constantly on the move, not only geographically (between places, projects, and political boundaries), but also socially (between people, communities, and organizations) and cognitively (between ideas, habits, and cultures) (Boltanski and Chiapello 2005). The majority of people who participate in these environments find tremendous value in the capability of such technologies for staying informed and in touch, forging new connections, and maintaining old ones. Most often, they do not think of their activities in terms of economic value, to themselves or to others. Such inattention is partly because of the convenience and feelings of connection that networking technologies provide, and partly because of the invisibility of the mechanisms. The contribution, as Boltanski and Chiapello have noted about computer-mediated networks in general, "must at once possess limited visibility ... and have meager value ... while contributing to [an enterprise's] enrichment" (ibid., p. 361). Modern persons find value in the use of social media technology through the rewards that it provides, thereby generating "enrichment" for the companies that provide the services.

5.4.1 Self-Validation as Reward

The precarity of current socioeconomic circumstances has increased status anxiety and the desire for affirmation from peers and others. This desire is met by the provision of the many microvalidation mechanisms provided by social networking technologies: "like" on Facebook, "follow" on Twitter, "endorse" on LinkedIn, "upvote" on Reddit, and so forth. Despite differences in style—for instance, between the "friendly" character of Facebook likes and the professional look of LinkedIn endorsements—these mechanisms share the purpose of providing peer and social validation. Displacing the common mechanism of "examination," which Foucault identified as a key disciplinary technique of modern societies, microvalidations draw on a combination of memory and emotion to encourage the conscious release and sharing of personal information (van Dijk 2013).

Professionally oriented sites such as LinkedIn and ResearchGate use the same mechanism but with an extra layer of formality, providing users with "profile stats," an "RG score," and other similar measures. The conscious and voluntary appearance of these behaviors has led some commentators to conclude that:

> [N]orms are changing, with confidentiality giving way to openness. Participating in YouTube ... Flickr, and other elements of modern digital society means giving up some privacy, yet millions of people are willing to make that trade-off every day. (McCullagh 2010)

The line between open and confidential or conscious and unconscious contribution, however, becomes murky if we take into account the effect of the delicate mechanisms of incitement employed by the sites. The People You May Know (PYMK) feature on LinkedIn, for instance, nudges people toward new connections, which are then reinforced by "invitations" from people one might not even know. While a person's response to such invitations seems conscious, refusal may feel awkward, given that turning down an invitation always carries a cost in guilt or feelings of lost opportunity.

5.4.2 Self-Promotion as Reward

The microvalidations offered by Web 2.0 technologies are enhanced by the provision of a space for self-promotion (a boon to all of us in a world of precarity in employment). The working of this mechanism is most visible for people with celebrity status (pop artists, actors, politicians) who depend on staying in front of people for their livelihoods, and have found in this space new ways to enhance and promote their image and to reach out constantly to their audiences. A wide range of such individuals, from Madonna to the Pope to the leader of ISIS to Malala Yousefzai, makes extensive use of these media to deliver their messages, mobilize supporters, and raise issues. But the average user can also employ the same mechanism to establish an intended image as a potential employee, connection, lover, friend, family-member-who-stays-in-touch. The "social capital" that is garnered can sometimes be further leveraged to acquire economic benefits. Some sites have, in fact, introduced incentivizing techniques—e.g., individuals with high-traffic profiles can post promotional materials on their Facebook walls on behalf of companies. Services such as Klout formalize this practice through aggregate analytics—i.e., by integrating individual information across social media platforms. The overall effect is the ongoing activation of the "cultural circuit of capitalism" (Thrift 2005, p.6), kept alive by the vast material infrastructure of data analytics, networks, and marketing techniques. In such an environment, personal tastes, thoughts, and profiles become commodities (Beer 2009).

5.5 On the "Fairness" of Communicative Technologies

From a heteromation perspective, the activities of users in social media environments count as labor to the extent that they (1) replace or support the work of paid employees (human resources, market researchers, product developers, corporate strategists, and others); (2) contribute to the bottom line (profit) of corporations; and (3) are not compensated financially in direct or indirect ways, or are compensated meagerly. The average person, separated from communities and individuals, driven by the desire to connect, willingly participates in these platforms to benefit from their networking capabilities. In so doing, they provide a trove of data that can then be used to structure their choices, regulate their behaviors, and provide or deprive them of opportunities, as the case may be. This arrangement might seem "fair," given the rewards the platforms provide for the average user, but is it? While this is a legitimate question from a moral perspective, it occludes the workings of the system, turning the question of human labor into a taken-for-granted supposition rather than as an object of inquiry. Our argument here seeks to avoid this trap, and the perspective of heteromation provides a way out.

This perspective brings out the asymmetries of the participation in networks as an uneven playing field. Participation enables individuals to connect, while at the same time depriving them of the "network-making power" of corporations, businesses, and the state (Castells 2009). Only a small number are able to leverage the capabilities of the networks for the purpose of data collection, turning data into a valuable asset. This group largely involves the *nouveau riche* of the internet age—that is, the founders of high-tech and social media companies that trade personal data (Forbes 2013).

Conceptualizing user contribution to online platforms as a form of labor highlights its invisible character, revealing the mechanisms of reward that drive people to consciously or unconsciously share personal information and other content. To blame people for the lapse of privacy or to applaud them for their openness tends to obscure these mechanisms, putting the blame or the credit on the individual. The oft-forgotten fact is that human beings, while enjoying a certain degree of choice and agency, are largely in the grip of social and personal predicaments, the impact of which is beyond their control. Meanwhile, the kind of self that emerges from the trade of data troves is, more often than not, an anxious, insecure, and uncertain self, vulnerable to the whims of data merchants who are on the hunt for "easy-to-sell audiences."

5.6 The Paradox of Mediation

Ferguson (2013) wrote of the tragic heroism of the lone individual facing the enormity of a continent, or a society, or a betrayal. Our loneliness, by comparison, lacks such grandeur. It is assuaged by a succession of fleeting moments of connection that momentarily quiet anxieties and insecurities. Within the difficulty of finding genuine connection online (or off), we accept "next-best-thing" substitutes. These substitutes are evident in, for example, services such as Google's Let's Play (LP) YouTube videos.

Let's Play is currently the most popular content on YouTube, with the bulk of YouTube's top channels affiliated with it in some way. Let's Play, and other amateur YouTube channels, are quickly overcoming mainstream media in viewership numbers, with millions of hours of content watched daily. The most popular YouTuber, PewDiePie, is an LPer with over 40 million subscribers. He earns millions of dollars a year in ad revenue.

Most LP creators, however, are ordinary people with only a few subscribers. (We discuss the bifurcation of networks into a small number of highly visible, successful participants alongside masses of ordinary

Figure 5.1
PewDiePie

participants in chapters 9 and 10.) LP viewers spend many hours a day watching recorded videos of (mostly) ordinary people playing video games in "playthroughs." Creators narrate their game action with a running stream of self-commentary and humorous banter. Creators themselves are also viewers. In an interview study of LP creators, one interviewee explained why he watched:

> I used to play games with friends when I was younger, but now I'm in college and my friends have moved away. Watching these people play games from my childhood and laughing and joking about it reminds me of that, and I can relive that through the internet. (Major and Nardi 2016)

Here, separation from childhood friends motivated viewing. LP creators tap into the angst of others' separation by designing video performances with speech, gestures, and settings crafted to project "authenticity" and "genuineness" (ibid.). The videos explicitly present the creator as the viewer's *friend*. One creator explained: "You've succeeded if your audience thinks of you as the friend next door" (ibid.). Viewers interviewed in the study stated they liked watching "just guys like me" and the "extremely down to earth, just regular people" found in LP videos. (ibid.)

While Facebook channels our communications by means of sophisticated algorithms written by PhDs, the amateurs over at Let's Play have taken on the possibly more difficult task of putting on a show in which they appear to like *you*! Creators report that most people who subscribe to and watch their channels are lonely or seeking connection with someone who shares their gaming interests (ibid.). The creator quoted above who guided his audience to think of him as the friend next door, regularly watched several LP channels himself. He spoke of the wistful, illusory quality of the experience:

> Often times I myself imagine myself having a great time playing [the popular online game] Minecraft with CaptianSparklez [sic] or Northern Lion [two prominent Minecraft LPers]. But since you can't actually do that, watching them play and make jokes is the next best thing.

He reports his experience without irony, cheerfully accepting its virtual, "next best thing" nature, much as the Yelp contributor mentioned in chapter 3 accepted Yelping alone on Friday night. People readily seek experience that seems, on its face, a little forlorn.

But we cannot dismiss this activity as mere spectacle, as passive viewing. Let's Play viewers are watching videos in which they are intimately familiar with the activity through their own participation in the games. And there is, crucially, some interaction with creators. Viewers may receive creator responses to comments they have posted on an LP channel, or responses to replies they have posted on a creator's Twitter account. Creators may acknowledge comments in a "shoutout" in a weekly vlog, a quite public airing of the connection between viewer and creator. Occasionally viewers even get to play with creators. Some creators devote episodes to playing games with fans whom they recruit on Twitter or in other social media. For viewers, this interaction is the pinnacle of Let's Play experience (ibid.), a chance to directly participate with the person they have come to think of as a friend. Several means of communication, though often infrequent and attenuated, bring viewers and creators together.

Other spectacles, by contrast, keep viewer and author/performer/creator strictly apart. Football, for example, a game played by only a small minority of men and by (essentially) no women, is nonetheless watched on television by millions of men and women. Football players are definitely not the friend next door, nor would they portray themselves as such. They certainly do not invite fans to play televised football games with them! The participatory nature of the internet, and its copious channels of communication, refuse the spectacle—viewers are swept into the (heteromated) action through direct and indirect interactions with creators. As noted in chapter 2, heteromation, and contemporary capitalism more broadly, work by *inclusion*, bringing the masses into the fold through varied forms of participation.

A central paradox of Let's Play is the odd sense of intimacy creators manage to establish with complete strangers. Creators are not the otherworldly Hollywood gods and goddesses of the era of massification, to be placed on pedestals and admired from afar. On the contrary, LP creators' appeal lies in appearing "authentic" and "genuine," even to the point of reminding viewers of childhood friends. If creators cannot project these qualities of authenticity, genuineness, and simple friendliness, they are doomed. One creator explained:

> Being genuine is the most important thing you can do. If you aren't genuine, they [the viewers] will know it, and they'll leave you for somebody else. (ibid.)

The threat of being left for somebody else intensifies as creators begin to attain higher subscriber numbers. Extending Let's Play-style friendship to lonely people becomes more difficult when creators cannot answer every comment or respond to every tweet as they did when subscriber size was small. Creators must manufacture "authenticity." Various manipulations come into play, such as the use of production techniques that make the videos appear amateurish. Ad revenue may be rolling in, but video content must not exhibit too much polish, even if the creator has moved on to using professional staff and high-end equipment. Unlike the perfection of, for example, the football replay, LPers may intentionally use low-quality techniques of audio mixing or footage editing, with scenes cutting off too early or too late, or images intentionally cropped poorly. One interviewee said:

> You want your content to look like you care, but you don't want it to look too good. Over-produced, over-edited stuff with what is obviously a team behind it seems less genuine. Once you're using boom mics and setting up light screens, how are you any different from IGN or Gamestop [professional gaming companies with YouTube channels]? You want to get as high quality as you can, but still within the scope of being amateur. (ibid.)

Even at the apex of Let's Play success, creators labor to crank out authenticity. For example, PewDiePie yells into his microphone in exaggeration when playing a game that might be scary or tense. Rather than formatting the audio to remove the static and popping that occurs when going beyond his microphone's volume threshold, these production errors are often retained (ibid.)

Another technique to keep the audience interested is disclosure of highly personal information. One creator said:

> I can understand why people get this weird feeling that they actually know you, when they're just watching your videos. First, you put out so much content, they're just way exposed to you. Second … you relate some personal stuff on there. You sometimes say stuff you wouldn't even say to your parents or real life friends, but here you are exposing yourself online. It's what people come to watch—that connection. (ibid.)

Both creators and viewers buy into a collective fiction of authenticity and connection. The audience knows that successful YouTubers employ teams of video editors and production managers, yet the suspension of disbelief that the videos are still somehow "amateur" is carried forward.

YouTube Let's Play videos leverage the heteromated labor of viewers and video creators, generating value for Google in millions of hours watched. A small number of LP creators turn their labor into compensated work, and a few become rich. Creators survive day to day on a treadmill of popularity they know they might easily fall off, well aware that their viewers are only too ready to leave them for somebody else if their performances fail to deliver a sense of connection. The paradox of spending hours a day watching videos when one is lonely, or separated from others with shared interests, instead of seeking face-to-face interaction, all the while understanding the manufactured nature of the mediated "connection" and "authenticity" does not bear easy explanation. Nonetheless, we can observe the outcome of the predicament of separation in the economic value generated through media such as Let's Play.

6
Cognitive Labor: A Story of Mental Toil

6.1 Introduction

In the contemporary imagination, computers are the epitome of thinking and intelligence. This image of machines is conjured through a combination of engineering marvels, metaphoric constructions, and discursive practices. A diverse group of actors—the technological elite, savvy entrepreneurs, the media, the state, the military, and a host of other enthusiastic practitioners and observers with a fascination toward technology—propagate this image in apparent unison, although often with divergent interests. The field of artificial intelligence has symbolized this phenomenon in a relentless fashion, with repeating cycles of fad, fanfare, and false starts interleaved with partial successes and fatal failures (Ekbia 2008). AI practitioners and enthusiasts have conceived and constructed systems with varying degrees of capability that mimic those of humans in one or another domain of behavior, but never in a way that encompasses overall human capacities. AI has kept alive one of the most ambitious dreams of modernity and beyond—namely, the creation of machines in the image of human beings. In so doing, however, it has also generated a great deal of confusion, obfuscation, and false expectations that have served, by and large, the interests of the embedding capitalist system and its dominant classes.

This chapter explores the relationship between capitalism and computing conceived as a domain that trades in "intelligent" machines. Whereas the previous chapter focused on computing as social practice, here we highlight computing as cognitive practice. Through a set of diverse examples—Amazon Mechanical Turk, the video game League of Legends, and self-service systems—we show how seemingly intelligent machines and automated systems need intense human involvement to operate effectively and deliver their promised functions. We examine how systems are set up to draw human cognitive labor into their functioning without acknowledging that labor.

6.2 Cost Saving: Value Extraction through Cognitive Labor

6.2.1 Self-Service Technologies: Cognitive Labor of Customers

A large number of technologies in a wide variety of industries belong to what comes under the rubric of self-service. Everyday life is increasingly populated by machines embellished with the label "automated": automated teller machines (ATMs) in banks, Automated Voice Response in insurance, retail, and customer support, automated check-in at airports, clinics, and hospitals, and automated checkout in supermarkets and fast food outlets. These applications and their devices and software even have their own industry magazine, *The Self-Service World* (www.kioskmarketplace.com). What entitles these machines to the label is their alleged capability to perform without, or in lieu of, a human being: a teller, a customer service representative, a cashier. The reality is more complicated. A close look reveals that a good part of the labor is, in fact, done by another group of human beings—the consumer, the end user, an intermediary such as a family member, or, in some cases, a new type of casual laborer. This labor costs capital little or nothing. In many cases, it replaces people who once had conventional jobs (with benefits and security), such as customer service agents at airports.

Here we examine ATMs and automated checkout systems.

6.2.1.1 Automated Teller Machines Banking provides a fairly straightforward case of the reallocation of tasks from paid workers to customers. ATMs, enabled by a vast infrastructure of secure networks, crediting systems, and transaction protocols, eliminate the human bank teller, saving banks money and providing a measure of convenience to customers. Online banking expands this logic by leveraging the connectivity of the internet to enable transactions without the need for physical presence. A more recent development seems to push the logic further through the introduction of the "ATM of the future," which processes more complex transactions than simple cash withdrawals or deposits. These new machines allow bank customers to perform functions such as cashing a check or interacting with a "virtual teller," who is a real person in a call center (NPR 2014)[1].

An industry analyst explains: "There is a very large need for banks to continue to control their costs, and of course, branches, branch personnel, etc., are one of the banks' largest costs" (ibid.). The scene of a bank equipped with self-service kiosks is not dissimilar to the notion of the "lights-out factory"—emptied of workers, this factory would need no lighting. The bank of the future *does* need lighting, however, because the *customers* are there working. Large-screen computers (rather than manufacturing robots) occupy the space (figure 6.1).

A proponent of the technology describes the "customer interaction model": "Tellers can walk over to the kiosk and show [customers] how to use it. By enabling and empowering the customer, we're enriching the interaction with the teller and the branch" (Bank Innovation 2012).[2] The rhetoric of "customer empowerment" obfuscates the fact that the customer has to do work that used to be done by bank employees, and that the "enrichment" goes to the bank in cost savings. So inured are we to the long lines pursuant to today's cost-cutting management practices that the bank's rhetoric unashamedly portrays the situation as a choice between bad and worse: "With self-service kiosks placed in teller lines, customers can interact with tellers for necessary functions, and perform self-service tasks while they wait. No one likes waiting in line" (ibid.). The coercive character of the arrangement is presented as an inevitable quandary, or even a favor being done on behalf of the customer.

Figure 6.1
The bank of the future (Bank Innovation 2012).

6.2.1.2 Interactive Voice Response Convenience, however, is not even a consideration in cases such as IVR. Every one of us has had the experience of interacting with voice systems that are frustrating because of their limited repertoire of responses or annoying because of the robotic delivery of individually recorded words strung together. Long menus with numerous choices, extraneous information (hours of operation, extension numbers), options that continue at each point in several languages, and incomprehensible voice prompts are some of the common features of these systems that impose cognitive burden on the caller who must navigate the system with patience and persistence. Despite the inconvenience, "Large and small businesses have adopted IVR technology because it saves money that would otherwise be spent on living, breathing (expensive) employees" (Roos 2015)[3].

While noticeable progress has been made in the area of speech comprehension in the last few years, with developers boasting "retail-like simplicity" of their systems (www.voxeo.com), studies show that frustration with these systems is still prevalent, running as high as 80 percent in the popular perception (Interactions 2012)[4]. People sometimes find workarounds to reduce the frustration (e.g., www.pleasepress1.com, gethuman.com), but these initiatives do not eliminate the cognitive burden on end users, who must deal with such systems on an ongoing basis.

6.2.1.3 Automated Checkouts A relatively recent addition to the world of automation, self-checkout kiosks have been on the rise in the last few years. In 2012 alone, 27,000 of these systems were installed in the United States. That IBM is involved in their manufacture says something about their growing significance.[5] Kiosks are currently adopted by a wide variety of service industries, from fast food to postal service to pharmacies. Market projections predict a $7.5 billion industry for automated pharmacies by 2018 (Huffman 2014).

The key concern of critics is focused on loss of jobs to these technologies, with the implicit assumption that machines are *replacing* humans. While job loss is real, what is missing from these analyses is that machines are not fully replacing humans, but rather reconfiguring the labor into the heteromated labor of end users who do the cognitive work. Since heteromated labor occurs on the margins of machines and organizations, often it is not even recognized *as* labor. Observant commentators note that many service occupations of the past no longer exist—"people who pump your gasoline, people who punch the button or your floor in an elevator ... clerk[s] selling candy and knick knacks at the hotel snack bar" (Huffman 2014). The commentators also notice that, "Today, consumers perform all those tasks for themselves and a lot more" (ibid.). As a matter of fact, consumers perform the tasks for corporations, not for themselves. The tasks are forms of heteromated labor costing individuals time and sometimes money (see Ritzer and Jurgenson 2010).

These examples illustrate the growing trend toward what is oxymoronically called "self-service" across various sectors and industries. These arrangements offer "service"—a word that means one person doing something for another—that is no service at all. People increasingly find themselves struggling with kiosks in stressful situations at airports, banks, hotels, clinics, and, very soon, in pharmacies. Business discourse does not recognize these actions as labor because corporations stand to benefit from hiding this relation. Their own explanations to journalists, however, often recognize the labor, e.g., in the comment about "expensive" employees, as well as the mention of Facebook's "unpaid workforce" (chapter 5).

6.2.2 Punish or Pardon: Cognitive Labor of End Users

We turn now to a different kind of cognitive labor—that performed by gamers who play the video game League of Legends. The coercive nature of the systems we have discussed so far in this chapter is absent—gamers willingly contribute labor to enhance their play experience.

Imagine you are a video game company and you receive tens of thousands of aggrieved emails *every day* from your player base of 67 million players worldwide. In 2011, this was the situation for Santa Monica–based Riot Games, one of the most successful video game companies in the world. In a perfect storm of technical affordances and demography, a large group of undisciplined players was spoiling the game for other players, and those other players were writing to Riot Games, the creator of League of Legends. The game is played in small teams of computer-matched players who generally do not know one another. They play short matches of about 30–60 minutes, and are unlikely to see one another again. Most "LoL" players are young men and boys, the demographic most likely to act outside conventional social norms, or even to be unaware of conventional social norms. LoL's mix of social and ludic factors created what

Riot Games calls "toxic behavior," including crude or hostile language, deliberately sabotaging matches, and displaying a bad attitude in an environment that is supposed to be fun. Players complained about toxic behavior, declaring that it was ruining their game (Kou and Nardi 2013).

Because multiplayer games demand the presence of other players as functioning parts of the system (Ekbia and Nardi 2012), if players leave the game, there is no game. People are readily induced to participate online, but, once there, they are not always docile and compliant. Game companies must find ways to govern the unruly, or their systems will fail. Riot Games thus instituted a form of heteromated labor in a system called "The Tribunal." This system allowed players to report disruptive players after a match through an online form. If a player was reported frequently, and by multiple players, the automated system prepared a case that considered the number and nature of the complaints. Players could log into The Tribunal and judge cases by examining direct evidence, such as chatlogs. Judges chose "Punish" or "Pardon" based on their assessment of the case. If the majority voted to Punish, the Tribunal suspended the player's account for varying periods depending on the pattern of infractions, possibly ending in a ban from the game in the case of persistently toxic behavior (ibid.).

The Tribunal apparently mitigated at least some of the behavioral problems (Senior 2011), and was declared a success. It is not in use at the time of this writing because after three years, Riot felt it had established positive community norms and improved player behavior significantly.[6] A designer we interviewed said, "One of the core philosophies of The Tribunal is to engage and collaborate with the community to try to solve player behavior together" (Kou and Nardi 2014). Drawing free labor from players reduced the costs of player governance and served to bind players even more closely to the game through the increased investment of their active participation. While the cognitive labor required of the Tribunal was willingly given, its economic value was hidden by high-touch words and phrases such as "collaborate," "community," and "solve problems together."

Such collaboration is arguably a positive development. We note, however, that in cases where gaming companies disagree with the player community, they have not hesitated to take legal action against player interests and/or to use their control of the software itself to enforce their choices, despite players' contributions of heteromated labor (see Nardi 2010, Postigo 2010, Kow and Nardi 2012, and chapter 7 in this book). Riot Games motivated players to do the work of establishing community norms and putting undisciplined players on notice, while reaping economic benefits, and yet true shared governance has not been tested. Judging by other cases, such governance is far from a sure thing.

6.2.3 Amazon Mechanical Turk: Cognitive Labor of Casual Employees

Not all heteromated cognitive labor is performed by consumers or players motivated by a spirit of participation. A different kind of motivation drives people to perform cognitive tasks in online labor markets such as AMT, TextBroker, and CrowdFlower. "Turkers" on AMT, for instance, enter into a contractual relationship with "requesters" to perform "Human Intelligence Tasks (HITs)" of varying difficulty. The requesters, who are often businesses but also academic researchers and other professionals, post tasks such as choosing the best among several photographs of a storefront, writing product descriptions, or responding to a survey. Turkers are paid by requesters according to the complexity of the task, as well as their own history, experience, and status. Payments vary accordingly, averaging between $2 and $3 per hour for basic tasks to $6 to $12 per hour for more complex tasks. Perhaps more important, hourly rates depend on how fast Turkers work and the acuity with which they select tasks. An elite group has become adept at not only working at speed, but at knowing how to choose tasks that can be completed quickly and that are put out by requesters with the best returns. A majority of this group lives in the United States, debunking claims about the "Third-World" weight of online labor (Silberman 2015).

Statistical information on the overall contribution of this type of labor to the economy is unavailable, but its value for requesters is hardly disputable, whether we consider businesses, researchers, or others. Studies have shown the overall quality of the work produced by Turkers for academic research is high, despite concerns over self-selection bias and other methodological issues (Sprouse 2011). (This point is disputed; see also Paolacci and Chandler 2014.) The bigger beneficiaries of the online labor market, however, are businesses, who manage microworkers at very low cost and without the contractual obligations of formal employment, including health insurance, retirement benefits, workers' compensation, sick leave, and vacation pay. In fact, the impersonal and algorithmic character of AMT and other mediators in the online labor market has arguably turned these laborers into "second-class citizens" (Silberman 2015).

6.3   The Invisible Value of Human Cognitive Capacities

While League of Legends players are incited to participate in heteromated labor to make their play space more appealing, many Turkers (and other online laborers) are driven by economic incentives. Some are meeting basic economic needs, while others seek small bits of money for discretionary purchases (Jiang, Wagner, and Nardi 2015). Silberman is very clear that some Turkers perform microwork "not by choice but because they are unable to secure other employment" (Silberman 2015, p. 14; see also Martin et al. 2014; Jiang, Wagner, and Nardi 2015). That these workers must work for multiple requesters distorts the true nature of the relationship as employment. The mediated character of the relationship enacted through the application programming interface (API) reinforces the distortion, allowing requesters "to manage human workers through software, as if workers were themselves software rather than people" (Silberman 2015, p. 20). All of this leads to the notion that what is really happening is computation, not labor—a notion that finds an unfortunate resonance in the new area of research called "human computation." The person and the person's labor disappear; only the output—the computation—is present, revealing once again the marginal character of persons performing heteromated labor. The machine, as the defining object, stands tall, and the human "steps to the side," as Marx said.

In between gamers and Turkers, we have the large group of consumers who increasingly find themselves caught in the grip of heteromated systems, without much recourse to alternatives. Pundits argue that consumers favor self-checkout systems because of speed, convenience, ease of use, and privacy. Surveys by business outlets claim that 85 percent of American and Canadian consumers are more likely to do business with a store that offers self-service[7]. A 2011 study of self-service in the hospitality industry reports that "consumers are not willing to use self-service technology in both lodging and food service environments, but 55 percent of those surveyed would be more likely to visit a hotel that offers the options to check-in/out" (Hospitality Industry 2011). To the extent that these reports are accurate, the fact of the matter is that consumers are intricately nudged toward certain types of preferences, or as one business analyst put it, "Self-service actually reinvents the customer experience" (Hospitality Industry 2009). The survey, for instance, shows that consumer preferences are largely driven by shorter waiting lines, faster service, accuracy, and privacy. While more than 70 percent of people surveyed mentioned these elements as motivators, only 40 percent favored "no interaction with clerk" in hotels and restaurants as reasons for considering self-service technology. In answering these questions, consumers may be thinking about the "long lines that no one likes" when choosing automated options, rather than formulating the question as one of a choice between a capable human vs. a machine. Business literature, however, tends to twist this, portraying choices as "interest in the technology and the belief in its potential for improving service" (ibid., p. 4).[8] One only need read TripAdvisor and Yelp reviews to see how much reviewers appreciate, and remark on, good customer service, informing other potential customers with this critical information for their choices.

6.4 A Paradox of Cognitive Capacity

One of the promises of technology is that it will free us for challenging, creative activity while the machines take over the menial labor. This vision was put forward in strong form in the era of augmentation, as we discussed in chapter 1. The examples in this chapter show that now we are veering off this path to a significant degree, and it is often the humans who end up performing the menial tasks. The increasing use of self-service devices and algorithmic management in systems such as Mechanical Turk marginalize humans, whose intelligence and creativity are sidelined. These systems ask mostly that we be patient and carry out the lockstep actions they present for us to perform. We conduct this labor because we must (to complete a task or for meager compensation), or because we agree to donate time in the absence of human mediators (e.g., LoL players). Sometimes we do so because it is more convenient when there is no human assistance.

What kind of a trade-off is this? Does our participation in such labor change who we are as social beings? Some people prefer to check their own groceries, to trust in an algorithmic system to discipline others, to order their food at a kiosk in a fast food restaurant. Humans can be unfriendly, unpredictable, unreliable—even toxic! The more alienated we become from one another (in part due to separation, precarity, futility, and monotony), the less socially attractive we become. Perhaps a self-reinforcing cycle is emerging, driving us toward increasing ease with machines, turning our behavior more machine-like all the while. Has anyone asked us—the average shopper, customer, user, player—whether or not we favor this change?

7
Creative Labor: A Story of Mental Magic

7.1 Introduction

Creative labor drives all human culture. Able to envision new material and social forms, humans have devised cultural solutions to the core problems of survival in every terrain on earth, a feat unequaled in the animal kingdom. Compared to our nearest primate relatives, with whom we share more than 90 percent of our DNA, humans are imposingly successful as a species, while those other primates are nearly extinct. The difference is attributable to the adaptation of culture, with its boundless capacity to stimulate and organize creative activity. Capitalism has not failed to notice this astonishing evolutionary development, and heteromation extracts sophisticated outputs from the creative labor of a multitude of ingenious humans.

Let us consider an early instance of this creative labor. In 1962, at the dawn of networked computing, engineers at MIT created the multiplayer video game SpaceWar! It was enjoyed by an emerging class of engineers and scientists at universities and think tanks. Members of these groups were constructing identities as creative, playful, geeky professionals (Golding 2015), differentiating themselves and their interests from the conventional masses (whom we discussed in chapter 3).

The contributions of these playful persons and their Cold War–themed game helped set the stage for the coming interest in video games and, subsequently, the vast and lucrative gaming industry. Supported by military funding, gaming became serious business. It currently generates billions of dollars, ahead of both film and music (Knoblauch 2015).

SpaceWar! was designed solely for fun. But industry noticed the game, seeing that it could be used to display the powers of the DEC PDP-1

Figure 7.1
Steve Russell programmed Spacewar!

computer, on which the game ran. This historic computer from the Digital Equipment Corporation (with a then-copious 4K memory!) received a marketing boost through the heteromation of the designers' and gamers' labor as marketers demoed the game to prospective customers to showcase their computer. In this chapter, we examine creative work that provides similar, contemporary sources of heteromated labor, namely, video game modding and graphic design. Here, heteromation is incited by the forging of identities and skills in specialist communities, contributing to the important project of developing personal life narratives.

7.2   Game Mods: Value Extraction through Creative Labor

Video games incorporate APIs for player-produced modifications ("mods") that enhance commercial versions of the games. Players use mods across a range of game genres, including first-person shooter, massively multiplayer role playing, strategy, survival horror, and others (Kücklich 2005; Taylor 2006a, 2006b; Kow and Nardi 2009, 2010b; Sotamaa 2009 ; Postigo 2003, 2010). Modders write the mods and upload them to host sites, where players download them.[1] The most popular mods have download counts in the millions. Modding picked up steam in the late 1990s, when id Software published the source code for Doom, leading to a profusion of new mods. The most famous mod evolved into the commercial game Counter-Strike, the best-selling game of its genre. Originally a mod of Half-Life, Counter-Strike set expectations that game APIs would allow modding.

Modders are rarely compensated. Some ask for donations or charge for premium versions. If a World of Warcraft modder receives five or six hundred dollars, that's considered a huge sum. Creating and maintaining a mod, though, may involve months of work (Kow and Nardi 2010b). The authors of an early suite of popular World of Warcraft mods wrote on their website:

> If you would be interested in working on some of the most used projects in World of Warcraft, please feel free to contact us...Unfortunately this isn't a "job" for us, it's what we do in our free time, so we aren't able to offer monetary compensation. (ctmod.net) (ibid.)

Figure 7.2

A player-created software modification for World of Warcraft showing players where to position their characters in difficult contests (Sindragosa's Lair is shown here).

Modders' labor is crucial to the success of video game companies. As Kücklich observed, "[T]he creativity of modders significantly reduces game developers' R&D and marketing costs" (ibid.). The gaming industry requires "constant creativity," as Kücklich (2005) put it, for inventing and reinventing games to keep gamers interested:

> Without the creativity of modders, developers would be hard-pressed to come up with new ideas, and it would prove hard to implement these ideas in the high-risk gaming market were it not for the huge "test-market" the modding community provides.

Companies can avoid expensive mistakes by keeping a close eye on modders' software, mitigating risk by choosing successful mods. Postigo (2003, p. 602) underscored the value of this free labor, remarking, "[Modding] manages to harness a skilled labor force for little or no initial cost and represents an emerging form of labor … on the Internet." Companies use the free labor of enthusiastic players to develop derivative works to license, thereby generating profit (Postigo 2010).[2]

In conducting participant-observation field research, in 2008, the second author and her coauthor, Yong Ming Kow, attended a dinner for modders at BlizzCon, Blizzard Entertainment's annual conference. Blizzard's UI Lead was in attendance. He told us that at the top of developers' concerns was good "gaming experience" (Kow and Nardi 2010a). This experience can be enhanced by incorporating mods, or certain features of mods, into the commercial product (ibid.). (Blizzard rewrites the code using these ideas, which cannot be copyrighted, to improve the game.)

Although such appropriation sounds exploitative (in an affective sense), it is a badge of honor for modders. Having one's mod brought into a game is a particularly desirable means of gaining recognition within the specialist community of modders and the Blizzard employees who interface with them (Kow and Nardi 2010a). The event we attended, at which modders gathered at Blizzard's own conference to meet with each other and with members of Blizzard's team, belied any feelings of exploitative appropriation. The genial evening was full of good cheer and interesting conversation among a rarified group of creative people who shared a passion for gaming and enjoyed membership in a community that understood each other's expert, even arcane, knowledge and skills. Modders were afforded the opportunity to interact with employees whose careers were aspirational for them or whom they simply deeply admired. Postigo (2003, p. 601) likewise observed that:

> [O]wnership of the productive process ... is what makes ... workers non-alienated from their work, and I believe it is the same ... process that compels modders to work hard for and identify with their labor.

7.3 Let Them Design Logos: Creative Labor in Graphic Design

Gaming is, of course, an entirely voluntary, discretionary activity. What happens to creative labor when earning a living is at stake? Schmidt's study (2013a, 2013b) of the "Let Them Design Logos" realities, as he calls them, of online graphic design contests examines heteromated labor in three platforms: i99Designs.com, CrowdSpring.com, and Design-Crowd.com. To build up portfolios, designers must submit designs to many contests. A designer submits an entry when a customer requests a design for, say, a logo. The platform posts the request and collects designs—often hundreds of them—submitted for free by designers hoping their design will be selected (i.e., it will win the contest). The customer chooses a design she likes, and only the designer of the winning entry is compensated. Schmidt (2013a) sardonically observed, "While the crowd is, by definition, not limited to a certain number, the money that is being paid out certainly is."

Schmidt (2013a) noted that unlike Amazon Mechanical Turk, where tasks can be split into small units, a graphic design must be holistically conceived and rendered as a single unit. Small payments, such as those for Human Intelligence Tasks, are not viable. The work is thus organized into contests for cost savings to requesters. The platform also takes its cut, receiving 40–45 percent of the payment. Hourly rates for designers end up being quite low because usually the designer receives nothing. Schmidt (2013a) provides a typical example:

> From the initial $300 that a client is paying for a logo contest, the platform takes off $120 right away. The client gets on average 116 logos, which leaves the designers with a chance of 1 in 116 to eventually get paid $180.

Schmidt (2013b) concludes that this "is certainly not a sustainable business model for the participating designers, but a bargain for the companies." The difference between the organization of labor in Mechanical Turk and design contests is an example of how flexible heteromation is.

Computer-mediated forms of managing labor in networks appear to provide efficiencies in bringing together employers and workers. But Schmidt (2013b) notes that the contest model leads to a waste of graphic designers' time, as well as wasted output. There is no use value for a logo designed for someone else. While the market promises efficiencies, they are often realized in asymmetric fashion, as one side assumes all the risk, producing efficiencies only for the other side. This asymmetry constitutes a hidden cost not generally recognized in economic models.

7.4 Rewards: Identity, Community, and Labors of Love

The creative labor harnessed in computer-mediated networks is not the immiserated, alienated labor of the worst of the Mechanical Turk HITs—in fact, the creative labor of modders and graphic designers springs from the excitement and devotion of people doing exactly what they want to be doing. Play is its own reward, as play theorists have long pointed out (Huizinga 1950; Callois 1961). For modders, the rewards of play tend to push aside economic considerations. One of the paid administrators of a commercial mod hosting site we interviewed said:

> I think that for the most part ... it's sort of a labor of love. It's a hobby. A lot of [modders] have donation buttons up, so if you want, you can send them ten bucks ... in PayPal or whatever, but very few require payments. Most of them, they just love [modding]. (Kow and Nardi 2010b)

Targett et al. (2012) reported that World of Warcraft modders interacted with Blizzard employees through online forums, becoming known to the employees and enjoying the recognition and status of employee responses to their posts (similar to what we observed at the modders' dinner). Identities as capable modders, skilled at design and the production of usable code, were important to the modders (ibid.). Postigo (2003, p. 599) noted that a critical motivation for modding was to experience a "sense of community." We observed this very sense of community at BlizzCon 2008, spending a day at Disneyland with the modders we were studying. A tight, sociable group, they had gathered together for their annual visit to an iconic venue of play and imagination, sustaining and solidifying their community.

We find the same satisfactions of membership and participation in creative specialist communities throughout the world of gaming—indeed, a huge draw of gaming is the possibility to shine without having to pass through formal educational barriers or professional certifications. Video gaming is as close to a level playing field as there is,[3] offering gamers the chance to create many forms of collateral materials and experiences, some of which are excellent. Gamers write game-based fiction. They produce machinima videos incorporating original music and clever storylines, sometimes garnering thousands or even millions of downloads. They create the Let's Play videos discussed in chapter 5. Others live stream their games. Those with a technical bent reverse engineer game mechanics to produce "theorycrafting" analyses avidly consumed by ordinary gamers. Theorycrafters post dense mathematical and logical expositions on forums so carefully managed that moderators set strict rules of participation, including the need to use proper English. (See Nardi 2010; Paul 2011; Choontanom and Nardi 2012; Bullard 2013 on theorycrafting.)

For the graphic designers, the rewards of heteromated labor are less straightforward. Designers enjoy their work as much as modders enjoy theirs, but the designers are attempting to earn their livelihood. They have graduated from college or from institutes of design (sometimes taking on significant debt). Economic precarity pushes them to continue to establish skills and credibility beyond the certifications of professional institutions.

Figure 7.3
We snap a picture of modders at Disneyland, BlizzCon 2008. *Source:* authors.

The predicament of precarity is especially clear here—expensive training is not enough to ensure paid work.

Capitalism thus sends people into longer and longer periods of unemployment and underemployment, drawing on a global cache of both workers and customers (Buchsbaum 2016). Those with low pay spend less money, but because global markets are efficiently linked through digital technology, enterprises find customers. This arrangement seems unsustainable in the long run, but for now, it enables companies to generate profit. Schmidt (2013a) points out the double whammy of globalization for graphic designers: they are not needed as customers, so their pay can be low, and they compete for work in a global market with designers from all over the world, some from very-low-wage countries.

A competitive spirit draws designers into contests. They believe they can "do better than the next person and therefore have better chances to win the prize" (Schmidt 2013b). Identities as expert designers playing a strong creative game ignite desires to win and to participate in the contests, even when the odds say they should do otherwise. The graphic designers are ideal workers in today's economy: entrepreneurial, willing to accept high risk, flexible, educated, competitive, creative. These traits help capital hide the value of designers' heteromated labor, suggesting that those who are "better than the next person" will prevail.

Contemporary capitalism is adept at stimulating activity in which we willingly engage, so that we tend to discount, ignore, or fail to notice the value of our labor. Here we have a manifestation of capitalism's own creativity. It continually finds new sources of value in labor, new sources "outside itself" (as Marx said it must), and obscures the value of that labor in varied ways. In the case of modders and graphic designers, the compelling reward structure draws people into activity and stimulates high levels of productivity.

7.5 Paradoxes of Creativity: Muddles in Modding and Cheating with Logos

Although the rewards of modding are clear, the corporate presence problematizes what our interviewee referred to as a "hobby"—a term that fails to acknowledge modding as part of an arena in which profit is a driving force. Video game companies encourage players to mod, but as soon as a particular mod does not fit a corporation's notions of how their game should be played, the mod can be disabled, with no input from gamers. The so-called "community" is fragile, even somewhat illusory, as decisions are made within rationales shaped by the concerns and power of the enterprise.

In December 2006, for example, fairly early in World of Warcraft's history, Blizzard disabled many mods to which World of Warcraft players had become accustomed. One was "Decursive," a mod that automated repetitive actions. Blizzard felt it made play too easy and that such mods were changing the nature of the game, diminishing its challenge. The modding community had disrupted World of Warcraft, taking it in directions Blizzard deemed unsuitable. Blizzard responded to protect the core gaming experience according to its vision, and abruptly, the mod no longer worked (Nardi 2010).

In 2009, a more serious breach occurred when Blizzard issued an edict (with threat of legal action) that modders could not charge for premium versions of their mods, or solicit donations (Kow and Nardi 2010). Modders were outraged. Although most had never sought money for their mods, they did not think it was up to Blizzard to tell them what to do. As one modder said:

> For a long time, Blizzard made no attempt to influence the [modding] community in any way other than to maintain the sandbox we played in. The tacit agreement was that we could develop [mods] for them that would both help us have fun and help them make more money. (Kow and Nardi 2010b)

The dispute was eventually resolved through the commercial sites that distribute mods, but feelings had been bruised. Modders responded with disillusionment. One forum poster wrote:

> The author community collectively believed that Blizzard was, well, sane, and didn't actually believe they owned everything. Now that they're threatening to sue over it, this assumption no longer holds, if it ever did. (Kow and Nardi 2010b)

Postigo (2010) observed that game companies often provide modders with tools designed to channel mod production in directions advantageous to the companies. He suggested that we should reflect on the fact that, "When we are invited to participate with tools made by others we ought to ask how our contributions are shaped through [the tool's] technological affordances." In other words, we should examine who benefits from such schemes. As van Dijk (2009) observed of the euphemism "prosumer," related terms like co-production, peer production, user-generated content, participatory culture, and human computation hide the economically valuable labor extracted in asymmetric relations of power. In a similar vein, Kücklich (2005) remarked that "ideologies" of collaboration veil "the precarious status of modders … disguis[ing] the power structures within which the modding community operates." The gaming industry "outsources" its own risk to the modding community, while portraying modding as a collaborative endeavor (Kücklich 2005). He called out Will Wright, the designer of the successful The Sims franchise, who stated that gaming is "a very collaborative process between the game developers and the players …" Kücklich responded dryly: "This statement is hardly surprising, considering the fact that The Sims has profited immensely from player-created content" (ibid.).

"Collaboration" does not extend to collaborative decision making for modders (or players); decisions occur strictly pursuant to corporate interests. These realities shape a larger paradox: lack of player governance is coupled with the warm friendships and mutual recognition that may develop between modders and game company employees, and the sincere approval employees may express toward modders.

Schmidt's discussion of design contests recalls Kücklich's and Postigo's wariness of corporate control. Schmidt (2013b) said, "While the crowdsourcing platform always wins, all the risk, liability, and workload has to be carried by the contributing designers." As a designer himself, Schmidt conducted participant-observation research in which he scrutinized the contest crowdsourcing platforms, assessing designer motives and actions. He reported:

> I can state that these websites are geared in a way that entices designers to put in far more hours than would be economically sensible. An important reason for this is the individual design portfolios that automatically build up on these platforms over time. … [T]he design portfolios stick with the online persona of the designer and can aesthetically be judged by anyone also much later. Designers tend to be very aware of this exposure and are therefore inclined to invest even more time. (ibid.)

The gap between reward and effort is exploited by a minority of graphic designers who are aware of the poor monetary returns and refuse to accept them. Instead they cheat:

> Others, who seem to be doing the math, frequently steal ideas and even existing designs from elsewhere on the internet to produce competitive work within a few minutes. This ... frustrates those putting in the extra hours and they in turn try to reveal the cheaters to the client. The exploitative mechanism of these competitive crowdsourcing sites for creative tasks therefore creates a poisonous and ugly working climate. (ibid.)

The ugly climate has given rise to initiatives such as No!Spec (no-spec.com), which urge designers to refrain from participating in commercial crowdsourcing contests (ibid.). It's an uphill battle though, because of the need for the portfolios. The number of contest participants is growing (ibid.). Still, No!Spec is an important site of resistance, featuring posts with titles such as "Refuse to work for nothing" and "Donating the unpaid labor of others." A YouTube video, "What Is Spec Work?" offers a witty critique of the contests (and is itself a display of the kind of creativity enterprises harness through heteromated labor).[4]

The heteromation of creative labor yields self-expression, fun, improved skills, identity development, a sense of community, and wider horizons. But it also entails possible disillusionment, disempowerment, and the poisoning of work environments. The graphic designers acted from economic precarity, struggling to marry creative impulses with the necessity of earning a living in a field of keen competition. Modders had more leeway, but were sometimes disappointed in the ways corporations treated them.

8
Emotional Labor: A Story of Caring

8.1 Introduction

The human capacity to care is distant from the aptitudes of machines. As designed objects created from human desires and constructed from inanimate materials, machines sit outside the struggle to survive. The existential condition of even the simplest organism centers on survival, the origin of the most primitive expression of caring. Marilynne Robinson captures living beings' capacity to care when she says in her novel *Housekeeping*: "And there is no living creature, though the whims of eons have put its eyes on boggling stalks and clamped it in a carapace, diminished it to a pinpoint and given it a taste for mud and stuck it down a well or hid it under a stone, but that creature will live on if it can" (1980). Artificial life might someday break the barrier separating machines from living creatures, but for the foreseeable future, it is likely that the work of caring will continue to require life.[1] This chapter considers two pertinent cases. The first is the deployment of social robots in eldercare (Ekbia, Nardi, and Šabanović 2015), and the second a form of banking that serves the poor in the remote rural areas and urban *favelas* of Brazil (Diniz 2007; Bailey, Diniz, and Scholler 2014; Bailey et al. 2016).

8.2 Social Robots

Industrial robots have been in use since the 1960s in various forms. Market saturation eventually reduced demand for manufacturing and operations robots, and governments in the United States, Japan, and Europe began to seek other application areas. The opportunities have, inevitably, been in less structured human environments, including homes (Gates

2007; Sung et al. 2007), offices (Hüttenraunch et al. 2004; Tsui et al. 2011), schools (Kanda et al. 2007; Tanaka et al. 2007), malls (Kanda et al. 2009), hospitals (Mutlu and Forlizzi 2008), and eldercare facilities (Broekens, Heerink, and Rosendal 2009; Inoue, Wada, and Ito 2010, Chang, Šabanović, and Huber 2014). The development of "social" robots for these environments brought out existing problems in robotics and AI regarding machine function under conditions marked by significant uncertainty in dynamic environments.

One way to address such conditions would be to precisely specify the tasks a robot does. But robots have generally been envisioned as more general-purpose machines—an attribute that would be undermined by task-specific designs. Another proposed solution is instrumenting human environments so that they are more habitable and comprehensible for robots (e.g., placing fiducial markers in strategic spots, RFID tags on objects, using ramps rather than stairs). The potential popularity or practicality of this approach is, however, questionable, since it involves both the costs of retrofitting existing spaces, and design choices that may not fit human preferences (Auger 2014). A third solution that is increasingly being pursued is to design collaborative arrangements for humans and robots that will allow robots to function successfully in human environments, and provide services that are beyond the capabilities of current technical know-how alone. In other words, the solution is to heteromate robotic labor.

One important source of heteromated labor for robots is for users to become involved in the physical labor that enables robots to function correctly in different environments. Forlizzi and DiSalvo (2006), for instance, found that the commercial robot vacuum Roomba required owners to do significant work to set up the environment so the Roomba could do its job well. With the Roomba and similar devices such as the Neato, it is necessary to remove certain furniture and articles from the floor, place barriers where the device might get stuck under a shelf or item of furniture, and be prepared to come to the rescue in areas of the home that seem to confuse the robot. These commercial applications of robots, as few as they currently are, are already producing heteromation in ways made invisible by the notion of "autonomy" that continues to accompany robotic products, carrying forward the captivating vision of autonomous machines that awed Phillip D. Dick's character O'Neill.

The infusion of more delicate human relations with robots brings up important questions, as we will see in the case of PARO, a seal-like robot used to treat elderly patients with dementia (Shibata 2012; Chang, Šabanović, and Huber 2013). One way this happens is that social robots actively request that people treat them in a social manner through the use of anthropomorphic and zoomorphic forms and naturalistic social cues (Turkle 2011). PARO, for example, is designed in the form of a baby seal, and looks and feels like a stuffed animal. It makes seal-like vocalizations and moves in response to human inputs. This design effectively hides the labor extracted from users by naturalizing the robot as part of a caretaking relation, subsuming it under the guise of a social relationship between the "animal" robot and the user.

Research on the social mechanisms of successful human-robot interaction has shown that robots are not as autonomous as roboticists popularly represent them to be. Their successful functioning involves significant efforts on the part of surrounding humans. The observational studies by Alač, Movellan, and Tanaka (2011) of human-robot interactions between children, teachers, researchers, and robots in a nursery school showed that robotic function is produced not through robots' technical capabilities alone, but through the social orientations and actions of the people around them. Vertesi's (2012) studies of NASA's Mars Rover missions demonstrated that the mission's robot was supported through collaborative forms of social organization in the science team and participants' embodied enactment of robots' activities.

8.2.1 PARO: The Caregiving Robot

PARO, the "baby seal," is designed to provide cognitive stimulation to elderly patients with dementia. Its soft plush surface is layered over a set of sensors and microphones that enable it to sense touch, sound, and changes in position such as hugging (figure 8.1). PARO responds to these signals through vocalization and movement. Several studies have demonstrated that use of PARO can improve patients' moods, decrease stress, and increase interactions with others (Wada et al. 2005, 2010; Shibata 2012).

8.2.2 PARO and the Division of Labor

Exactly how these outcomes are produced can be viewed in different ways. In some accounts, the autonomous robot produces the outcomes.

Figure 8.1
PARO, the robot seal. (Photo credit: Selma Šabanović)

In other accounts, human labor and the design of PARO together produce the outcomes.

In the first account in the literature and in robotic demonstrations, PARO is described as "autonomous." PARO's inventor, Takanori Shibata (2012, p. 2529), says that the robot

> acts autonomously ... while receiving stimulation from the environment, as with living organisms. Actions that manifest themselves during interactions with people can be interpreted as though PARO has a heart and feelings.

At other times, Shibata has said that PARO is an underdetermined and flexible artifact, as in this statement from a talk for the Japan Society in New York:

> PARO has a limited number of functions as a machine. But through the interaction with human beings I designed PARO to evoke associations in the human's mind. So in that case, it is not necessary for PARO to have all the functions, but the interaction can enlarge the number of functions. (Shibata 2007; see also Ekbia, Nardi, and Šabanović 2015)

This second way of describing PARO's interactivity suggests that people's orientations and behaviors toward the robot play a significant role in bringing forth the desired therapeutic and social effects. But evaluations of PARO have often been based on the first perspective, shaped around its putative autonomy. In this view, an autonomous PARO connects in unmediated fashion to an elderly person. Thus, it makes sense to measure the effects PARO produces, rather than analyzing the effects of a joint production of PARO *and* the social interactions that develop around it. The notion of the autonomous PARO informs most of the research in the literature. Studies typically measure health outcomes following interaction with PARO, such as changes in stress level, depression, and engagement with other people (see Chang, Šabanović, and Huber 2014 for a review). Changes are attributed solely to PARO.

In a key paper evaluating PARO, Wada et al. (2005) first explain the technical capabilities of PARO, then the subject population, and then the experiment they conducted. The experimental setup is described as follows (p. 2798):

> Two seal robots have been given to the elderly people at the facility on two days per week. We prepared a desk for the robots in the center of the table, and people were arranged up around it. They interacted with the robots for about one hour at a time. Since not all subjects could interact with the robots at the same time, we had them take turns for equal periods of time.

There is no mention of caregivers, yet PARO is always mediated by caregivers. The use of the passive voice in the description deletes caregivers' from the scenario—it is actually they who are sitting at the table with the elderly people, encouraging and managing the interactions. Later in the paper, the authors note that the caregivers liked PARO, recognizing their presence. But their invisible labor does not figure in any aspect of the evaluation.

Shibata's second perspective on PARO as a more interactively situated artifact suggests the need for a broader framing for study. Evaluation shold consider not only the effects of the technology, but how they are produced. Ethnographic studies paint a nuanced picture of the human labor involved in PARO's use, generating more complex analyses of the sociotechnical mechanisms producing outcomes (Chang, Šabanović, and Huber 2014).

These analyses consider individual variation in response to PARO, unpacking some of the complexity hiding in the aggregate numbers. For instance, Chang, Šabanović, and Huber (2014) reported the surprising finding that many patients ignored PARO. They did not become engaged until staff, family, or friends induced them to participate. Environmental factors controlled by staff, such as the placement of the robot, were critical. Unless caregivers placed PARO away from distractions competing for patients' attention (such as puzzles or television), the robot was not interesting to patients and they made no attempt to interact with it. These intricacies, crucial to understanding how PARO actually works, fall outside the bounds of what can be discovered using biomedical metrics such as aggregate changes in stress levels. Chang, Šabanović, and Huber (2014) summarized their observational study in a US nursing home, noting how their study differs from most research on PARO:

> Our results indicate that interaction with PARO in the public is not as immediate and straightforward to establish as might be expected from prior studies, which rarely discuss the mechanisms by which interaction is started and maintained, or from videos of older adults interacting with PARO on the robot's website (http://www.parorobots.com/video.asp), which present only successful interactions. We rarely observed interactions between older adults and the robot spontaneously occurring in the field; more often interactors did not notice or were hesitant to interact with PARO (especially for the first time), or their attention was divided between the robot and other interests such as TV, puzzles, or observing other people. We also saw that most successful interactions of older residents with PARO occurred when staff and family initiated and encouraged them.

Research that conceives evaluation as a set of quantifiable variables underspecifies the functioning of digital technologies, obviating knowledge of how the technologies work in real deployments and how they might do some good if properly used. Systems such as One Laptop per Child, for example, have fallen prey to similar misconstruals, squandering resources and leading to disappointment in failing to consider how the technologies operate in real practice (Cervantes et al. 2011). Evaluations framed as variables correlated and numbers crunched obscure the ways patients encounter a robot like PARO, and how humans contribute to robotic function.

Chang, Šabanović, and Huber (2013) provide an account of PARO in practice, demonstrating the essential guidance of a therapist in directing activity:

> The therapist in our study adopted multiple methods of generating interaction in the group, including asking questions, showing the interaction with

PARO, and directing participants' attention to others interacting with PARO, switching between different mediation methods based on the personal needs of each participant and the group dynamics.

Here, the emotional labor of caring involved calibrating the personal needs of each patient and tracking fluid group dynamics to release as much value from PARO as possible. Such care is the difference between a patient who turns away to gaze at the television, and one who interacts with PARO, gaining whatever benefits a social robot may produce.

In acknowledgment of the importance of human labor to PARO's functioning, Shibata's long-term collaborator, Kazuyoshi Wada (2010), developed guidelines for introducing and using PARO with older adults in institutional settings. This work gives recognition to the notion that robotic systems, their functions, and consequences should be developed, implemented, and studied as part of broader social systems, potentially creating a new form of expertise—the knowledge and ability to effectively implement assistive robotic technologies in human environments.

After considerable experience with PARO based on observing its use in multiple settings, Wada et al. (2010) softened claims of autonomy. They noted: "The effects of robot therapy using PARO emerge through the interactions among participants, PARO, and caregivers. The caregiver who manages the therapy plays an especially important role in achieving effective therapy" (ibid., p. 533). However, failures in PARO's use were attributed solely to caregivers' ineptness—not to PARO's design: "Thus the effects [of PARO] have been influenced by the varying skills of the caregivers" (ibid., p. 537). Varying effects, of course, have other potential sources, notably, varying patient responses to the design of the technology itself.

The text of Wada's guidelines (inadvertently) reveals the variability, affording a glimpse of the darker side of PARO. One guideline notes that a good caregiver can "reduce the [patient's] apprehension by showing PARO at her eye level." Another instructs that if a patient begins beating PARO, a correct response is a remark such as, "PARO says, 'Ouch! Please treat it tender.'" Another tells the caregiver to stimulate interest in a patient who is bored with PARO: "'PARO is cute, isn't it? Stroke it, please'" (Wada et al. 2010, pp. 533–534). Yet the larger, generalized claim about PARO is that its design evokes love: "[M]ost people love PARO, and are waiting to interact with PARO" (p. 534). Apprehension, violence, and boredom—described in the nitty-gritty of the guidelines—are left out of the story of the baby seal with the "pure white fur" (p. 533). While PARO's effects are acknowledged to emerge from "the interactions among participants, PARO, and caregivers," the design of PARO drops out of the equation of evaluation, reducing failures of patient engagement to caregivers' actions.

8.2.3   Memory: Value Extraction through Emotional Labor

The question of how PARO elicits the emotional labor of caretakers, relatives, and friends is a difficult one. One answer is that PARO's form naturally appeals to our playfulness. PARO looks like the most cuddly and adorable of all toys—the stuffed animal—an object we might respond to because of pleasant childhood associations. Turkle (2011) suggests that the form of such social robots returns us to childlike wonder, disarming our anxieties.

PARO's design also reflects, to some extent, what we can call the "attribution fallacy" (Ekbia 2012), in which people attribute far more meaning than is there to inanimate objects or strings of words. The most famous and enduring example is the Eliza program in which Weizenbaum's software system prompted people to interact with it as though it were a real person, sharing their secrets and unburdening themselves on sensitive matters (Ekbia 2008).

The dynamic of attribution undoubtedly comes into play with PARO. But the matter is complicated. Elderly patients may respond to the furry object, attributing life to it: "Sometimes they kissed [PARO] with a smile," Wada reported (2005, p. 2799).[2] But for caregivers, acts of attribution may be more instrumental than naive. Within the forbidding reality of caring for people incapacitated by dementia, so old they no longer have value to society, caregivers face challenges that would paralyze most of us. The institutional mix of guilt, fear, and a sincere desire to care for the most needy is deftly leveraged by the social robot. PARO provides one more tool for what is an almost impossible task. Rather than falsely attributing animacy to PARO, caregivers use the story of the baby seal to their occupational advantage. Deploying gentle, childish fictions that sound the note of the patients' dependency, but also the safety and innocence of childhood, caregivers leverage the form of the robot to stimulate memories that weave stories of caring and nurture. Thus, in a Japanese nursing home, caregivers created a "home" that the robots returned to at

Emotional Labor 137

the end of the day, to recharge their batteries and "sleep" (see Wada et al. 2005).

With its primary colors and childlike simplicity, the "home" has a kindergarten feel. Through the narrative of the animals returning home, the PARO units are integrated into the caring ethos of the nursing home. The robots themselves become vulnerable creatures to care for, allowed to slumber quietly as they enjoy their pacifiers. When PARO runs out of batteries, caregivers continue the narrative: "A caregiver who noticed the [dead battery] decided to pick up PARO to recharge it. The caregiver said to [the patient], 'I'll bring PARO to the bed, because it fell asleep'" (Wada et al. 2010, p. 535).

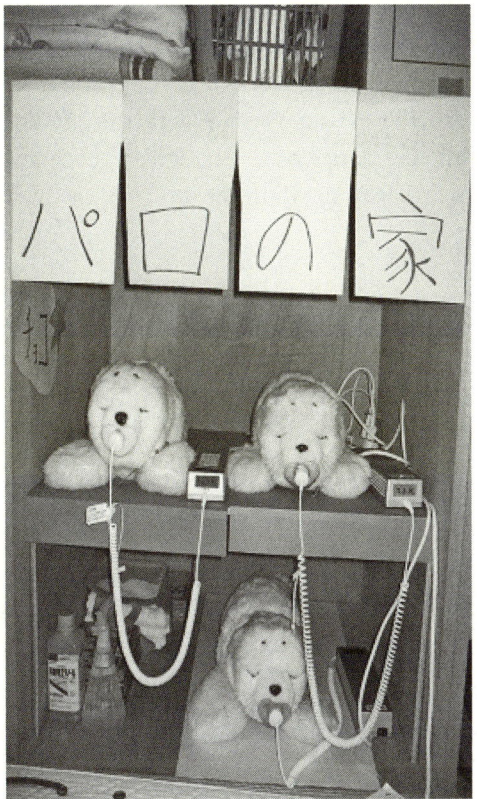

Figure 8.2

PARO units in a nursing home in Japan. Note the pacifiers and mats. (Photo credit: Selma Šabanović)

Patients themselves may feel that they are taking care of PARO, continuing the fiction in responding to what is basically a stuffed animal with some simple interactive moves. What makes all of this work is not PARO's "autonomy," but a design that gains purchase from simple human stories, stories that encourage people to read into the artifact attributes that it doesn't really have—in this case, "affect." Patients supply heteromated labor to invest PARO with meaning, once they are primed by the caretakers and their insistence on PARO's cuteness, cuddliness, and friendliness.[3]

Here we have dual forms of heteromated labor: the first is that of the caregivers, the second that of the patients. In more and less instrumental ways, caregivers played out the story of the baby animals to generate interest for patients. Tellingly, Wada et al. (2005, p. 2798) described how staff members "felt difficulty of communication with [patients] because of a lack of common topics among them." PARO gave them something to talk about, easing the rigors of the demanding work of dealing with incapacitated people. The patients, to the extent possible with their fading mental faculties, responded to the entreaties of the caregivers, engaging the robot as they were urged to, drawing on whatever scraps of memory remained to them.

8.2.4 Precarity and PARO

In our discussions of Amazon Mechanical Turk and social media in previous chapters, we had some financial facts and figures to work with such as payments for Human Intelligence Tasks, profits and market capitalization for Facebook, and other statistics. With social robots in nursing homes, the economic logic is different, and we have no direct data. Value is extracted by increasing the competiveness of nursing homes (improving care, offering the exciting feature of robots), while at the same time keeping wages flat. People accept difficult, ill-paid jobs in nursing homes because of financial precarity. Once there, they do what is asked to keep their jobs.

While caregivers have learned new high-tech skills, which they perform with finesse, institutions downplay their knowledge of assistive robotic technologies as simply a matter of "training." Caregivers are depicted as points of vulnerability in effective use of PARO: they "may not notice whether their actions are helpful or harmful," Wada et al. (2010, p. 534) comment (rather critically). Rather than receive better pay to match enhanced skills, caregivers are admonished to pay closer attention to the robot. New competencies and routines of managing the technology are neither acknowledged nor compensated. Caregivers deploy PARO with verve and imagination, bringing much more than attention to the job, but precarity of employment obviates the need for employers to take account of caregivers' new contributions.

The PARO technology might, in fact, offer benefits to caregivers. They used the technology to ease burdens of alienation, in particular "difficulty of communication with patients." PARO could make "the general atmosphere ... brighter" Wada et al. reported (2005, p. 2800), reducing the "gloomy" feel of nursing homes (p. 2798). The more important need for caregivers, however, is to hang onto a job in a situation of structural unemployment, guaranteeing their acceptance of the technology.

8.2.5 Hiding Value in the Wage

The economic value of the PARO technology is hidden through the ambiguous relationship of the existing wage to the real value of caregivers' labor. Without a union to specify work rules and pay grades, workers add new proficiencies and duties to their work, but have no mechanism for demanding, or even discussing, increased compensation. As in many forms of heteromation, the work tasks are small and not unpleasant, and their very meagerness fails to provoke feelings of disruption or undue interference. The tiny, value-producing moments are hidden in busy streams of ongoing activity.

Yet, there is real economic value in these social robots, in particular for the manufacturers. Wada et al. (2010) says, "The use of a therapeutic seal robot, PARO, in various facilities for the elderly is spreading around the world." Shibata (2012) examined the commercialization of the PARO technology, including regulatory approval in many countries. PARO units have been sold in Japan, Europe, and the United States. Heteromated labor that drives technological systems provides economic value to these privatized operations. Manufacturers of robots and owners of nursing homes and other institutions where robots might be deployed extract profit from caregivers' labor, as well as the labor of friends and family who try to help.

Emotional labor is not strictly skill based, and certainly not primarily a result of "training." It is, on the contrary, the result of a lifetime of being a human person: of memories that begin in childhood, and of the development of habits of observation, compassion, empathy, and sympathy. Heteromated labor appropriates the deepest, most hard-won human capabilities. There is no machine substitute. Could there ever be? Possibly, but the costs of developing such machine capacities make little economic sense, given the army of cheap human labor that already has them.[4]

8.3 Correspondent Banking

We turn now to another, very different, instance of emotional labor in an unlikely context: the banking industry. In 2001, Brazil's national government and selected private banks collaborated to increase financial inclusion for Brazil's poor through a novel point-of-service system called "correspondent" or "branchless" banking (Diniz 2007). Correspondent banking serves millions of Brazilians too remotely located to get to the bank, as well as urban poor in *favelas*. Merchants in small grocery stores, drugstores, lottery shops, and internet cafés are given computer equipment to enable local banking clients to conduct simple transactions (Bailey, Diniz, and Scholler 2014; Bailey et al. 2016). Clients can make deposits, withdrawals, and utility payments, and receive state benefits. Merchants adopt the new role of "correspondent banker," continuing their usual jobs, but also receiving small commissions for providing access to the banking equipment in their establishments.

The technology is simple, usually a kiosk or personal computer networked to a bank, and a few peripherals such as card readers. To conduct transactions, a client swipes a card at the kiosk or logs in, enters a personal identification number (PIN), navigates the user interface, and takes a receipt, which is given to the correspondent. The client then receives or deposits cash (e.g., receives government assistance or adds to a checking account).

This arrangement is a classic "self-service" setup. The stipulated actors and actions formed the use case intended by the system's designers. But Bailey et al. (2014, 2016) found that many clients were unable to use the technology. Even a simple keypad proved too difficult. Instead, clients handed the correspondent a small piece of paper on which their PIN was written. The correspondent looked at the paper, typed the number, and returned the paper to the client. The system was able to meet its government-mandated goal of wide financial inclusion not because of its self-service capacities, but because correspondents heteromated the system, operating the technology on behalf of clients who could not do so.

This case of heteromation would appear to be an example of simple cognitive labor. But the correspondent bankers came to occupy a role that exceeded task-based customer support. Over time, clients began to ask the correspondent bankers questions about bigger issues: long-term financial planning, savings, loans, starting businesses. Clients entrusted correspondent bankers with vital, private financial information, along with their hopes and dreams. Correspondents used this trust to open up the financial system to clients who could now learn of new services such as loans and savings accounts (Bailey et al. 2016). This change was profitable for the banks, increasing services sold (Diniz 2007).

Correspondent bankers performed another kind of emotional labor by helping clients intimidated by the whole idea of banks and banking. The correspondents personalized banking for these clients, caring for them, getting to know them, explaining how banking worked. One lottery store owner said:

> There are people that come here so often that I know their names, date of birth. Sometimes an elderly person forgets to bring a picture ID and it's a requirement from the bank and when they get here, they want to talk to me, because they know I remember. (Bailey et al. 2016)

Caring entered the impersonal technological apparatus the banks and the government had assembled, shifting and expanding what was conceived at the outset as the correspondents' role as proprietors of bank-owned kiosks.

8.3.1 Drivers: Reappropriating Resources

The economic precarity of millions of poor Brazilians drove the development of correspondent banking. But Diniz (2007, p. 13) points out that Brazil had been investing in sophisticated banking technology for decades: "Because of a series of historical circumstances, the Brazilian banking industry is widely acknowledged as one of the most advanced in the world in its use of information technology." The technology to increase profits was at hand, and the government–private sector partnership enabled correspondent banking to grow rapidly, reaching millions of people who had few or no banking services. There are four times as many sites of correspondent banking as traditional banks in Brazil (about 72,000 correspondent sites) (ibid., p. 17). The Brazilian correspondent banking model is unique "because of its reach and scale, the quality of the services provided, and the new technological platforms that have enabled these services" (Kumar et al. 2006, p. 10). As we saw with Amazon Mechanical Turk, there was no need for build-out or maintenance of expensive infrastructure—existing, rent-free facilities were appropriated. Diniz (2007, p. 18) observed, "The [banks] multiplied the number of service points without incurring the costs associated with the operation of traditional branches or mini-branches." Heteromation often appropriates existing means of production, without cost, whether Turkers' own personal computers and home workspaces, or the real estate of small shops for banking kiosks. Powerful, centralized means of production such as servers and banking software remain under the control of enterprises.

8.3.2 Value Extraction Mechanisms: More Social than Technical

Although heteromated labor is not directly included in the calculation in Diniz's analysis of bank profits, he indirectly acknowledges such labor, saying, "[L]ow-income groups [in Brazil] tend to shy away from traditional banking facilities which they consider hostile and unwelcoming. This is not the case with local retailers, where these same people feel at ease" (Diniz 2007). Bailey et al. (2016) talked to a copy store owner who said:

> I like [being a correspondent] because I can help somehow, sometimes. I meet people who are so simple, who haven't ever gone to a bank, or who are too shy to go there, or they send someone to get their pay checks. And I can visit them at their home, and have them come to my store, and I treat them like any other, as a public servant. I think this part of social equality is very important and very rewarding.

It seems unlikely that the massive uptake of correspondent banking would have occurred without the merchants' labor. Correspondent bankers began to see themselves not just as small business proprietors, but as guides for what they called "humble" people (Bailey et al. 2015). The copy store owner, for example, engaged in emotional labor as he calmed the shy, visited clients' homes, and asserted their dignity and equality.

Correspondent banking differs from some of the other cases we have considered, such as social media, with its elaborate technological disciplines for channeling interactions to package data for corporate buyers. Correspondent banking entails few choices and little flexibility. But to achieve even the simple basics with the Brazilian populations served, social mechanisms came into play. The complexities of the origins of the correspondents' emotional labor are beyond the scope of our analysis, but clearly, the development of a person is an intricate, lifelong project. Vygotsky (1986) spoke of the "higher psychological functions" that emerge as a result of engagement in sustained cultural activities. These functions are foundational to the development of ethical systems such as those than underpin notions of "social equality," which were crucial to the success of correspondent banking.

The case of PARO bears similarities. To use the robot effectively, caregivers marshaled capacities constructed over a lifetime, delving into childhood memories, summoning compassion for their decrepit patients, and thinking of ways to insert PARO into patients' lives with a playful creativity. Just as PARO was supposed to be autonomous, correspondent banking was supposed to be self-service.

8.3.3 Hiding Not Only the Economic Value But the Customers Too

We have argued that heteromated labor tends to be invisible, erasing the human contributions that allow technological systems to function and profits to be made. Bailey et al. (2014, p. 12) talked to bank employees in urban Brazilian banks who "were astounded to learn from us that clients shunned the simple interface of the keypad and entrusted their passcodes to correspondents." The bankers were completely unaware of the human labor required to help clients who could not operate the technology, as were the engineers who had designed the system. Bailey et al. (2015) reported that

> When we visited a third-party integrator in their São Paulo corporate offices, the engineers with whom we spoke were surprised when we demonstrated the position of the keypad and how clients handed over slips of paper for store owners to enter PINs. As designers of the ICT, these engineers had no sense of the technological timidity in remote and underserved areas. Banks were likewise oblivious to the correspondents' extra actions and the larger social role that correspondents came to play.

The labor required for the correspondent banking system to function escaped the notice of bank managers and the technical staff who had designed the system. In a deeper twist, the arrangements of correspondent banking hid the customers themselves. Just as the crowdworkers of

Mechanical Turk dissolved as people and became software in the eyes of managers ("as if workers were themselves software rather than people" [Silberman 2015, p. 20]), the impoverished Brazilian masses disappeared from the purview of urban banks once correspondent banking was established. The managers were happy that correspondent bankers served poor clients because it left them with a higher class of customer. One bank manager said:

> If … a lot of people don't have a bank account, don't have access to a bank, don't have proven income, they [still] have to pay their bills somewhere. And this public doesn't interest me. … The banking correspondent takes out this load of people, leaving my branch wonderfully beautiful, [leaving] not only those who have money, but those who want my money [for loans] … [Before correspondent banking], thousands of non-clients, pedestrians, would come, fight, speak loudly, and provoke arguments. (Bailey et al. 2014)

The bank manager described what he saw as ugly physical realities that correspondent banking had relieved him of: loud speech, fighting, arguments. He confusedly referred to "non-clients" and "pedestrians," yet the poor going to the bank were paying bills, a legitimate banking function in Brazil. The banker's problem of the "uninteresting" public was solved by correspondent banking, transforming the physical bank into a "wonderfully beautiful" place, once it was devoid of certain customers. The theme of physical beauty formerly marred by the masses echoes in the words of another banker who said:

> In 1999 a lot of people wanted to pay their bills, but they were not banking customers. Let's say 70% of the population … They build banks with marble; everything is fine, beautiful for one type of client, let's say a high-level client that is highly placed in society. And suddenly a lot of poor people come in to pay their bills. (ibid.)

By contrast, a correspondent banker in a post office pointed with pride to his role in achieving financial inclusion, underscoring the emotional labor that went into it:

> [T]he customer service of the post office is much more humanized in comparison to a bank. A bank is colder, they don't offer the service that we offer. It is good for the people. … (ibid.)

8.3.4 Rewards: Gaining through Service

With clients allowing correspondents to see their PINs and finding a supportive environment in which to use unfamiliar technology, clients' trust in correspondents grew. Correspondents began to provide services well beyond helping with PINs. One said:

> The most rewarding thing is the social equality. [Clients] feel that they are the same as anyone who lives in the city ... They can have payroll and loans, and so on. It doesn't matter if they earn the minimum wage. (ibid.)

A grocery store correspondent spoke of how he worked diligently to solve a range of problems for clients:

> This is a small town, and they [clients] expect that we know everything, things that they expect to resolve here, and [even things] we don't know. I always try to chase the answer for them. (ibid.)

Correspondent banking was an opportunity for merchants to become the benefactors of "humble" people—of the elderly, of people afraid to set foot in a bank. Correspondents were not completely selfless; they stood to gain increased traffic in their stores. But the dedication to service, along with their understanding of the social problems they could mitigate, at least a little, through their service, speak to more than self-interest. Correspondents filled gaps in the banking technology through emotional labor conceived as a mission informed by "social justice," or "social equality." They drew on their sense of what was good about themselves, and the growing admiration and trust of customers, to provide even more than the government had envisioned when it instituted correspondent banking.

8.4 A Paradox of Compassion

It seems likely that creating robots with sophisticated memories, compassion, and empathy, if such a thing were possible, would be too expensive to even contemplate. To develop a person, it takes a village, as we know, and it takes a village a very long time to produce even one unexceptional human being. PARO's simple, interactive moves required years and millions of dollars to achieve. It is perhaps no accident that PARO is installed, in its initial deployments, in locales of those whose declining faculties make them oblivious to the robot's scant competencies. In the West, robots are critical and pervasive in the military-industrial sector, but such robots don't need memories and emotions. It's easier to hire the labor of immigrants and the underclass for caring work in hospices, nursing homes, and those dark corners of patients' own homes, where many spend their final hours.

In Japan, however—a low-birth-rate, immigration-averse society—the cultural logic supports a turn to robotic solutions to manage the growing burdens of the increasing elderly population. Will eldercare robots become more sophisticated and caring? Or will society play an end game only it can win in which the least able, in fact, those literally ready to drop, will have no choice but to accept what is offered, even if it is not much better than PARO? We are reminded again of *Sleep Dealer*; when laborers wore out, they dropped out of sight.

The wider cultural imagination falters at the prospect of robots such as PARO, and we simply do not think of them. Hollywood's personifications of artificial intelligence as sexualized companions and terminators could not be further from the truth of today's robots designed for caring labor (or, for that matter, designed for the military-industrial sector). The heroic robots of Hollywood films are not only ethereally beautiful or possessed of super strength or intelligence, often they are trying to become human themselves, an increasingly narcissistic conceit. We are unlikely to see a drama about robots in nursing homes.

A country like Brazil, with its populous, youthful society stratified by strong class divisions, has little need of robots. The proficient, caring labor of the correspondent bankers, with their ideologies of equality and social justice, are, for the moment, a robust form of heteromation that provides a human touch to high-tech infrastructure. For eldercare, we might imagine a society that draws on the goodwill characteristic of those such as correspondent bankers, and adds a dash of technology to help with, literally, the heavy lifting, and other such tasks. In the Netherlands, for example, students can live rent free in nursing homes if they keep the elderly residents company.[5] How about throwing a few robots into the mix to help with certain tasks, leaving social interaction to the amiable Dutch students?

All of these examples seem to point to the same paradox—namely, the "optimal" choice between human caring labor and compassion on the one hand, and machines' limited but overblown capacities on the other. That the paradox plays itself out differently in different contexts—PARO vs. caregivers in Japan and the US, banking technology vs. correspondent bankers in Brazil, or students vs. potential robots in Netherlands—shouldn't blind us to the underlying similarities and their socioeconomic drivers.

9
Organizing Labor: A Story of Commitment

9.1 Introduction

Human labor is inherently collective, in the sense that individuals rely on others' labor in order to bring their own to fruition. This dependence calls for some kind of structure and organization to enable and support a division of labor within the collective. Organizing, however, incurs costs—of communication, coordination, control, and collaboration. These costs represent an interesting problem for organizations because they must expend a portion of their resources to manage the division of labor. The larger the number of people involved, the higher the cost of management.

Historically, societies differ in how they divide the labor and how they handle the costs. We have discussed how in feudal societies, the division of labor obligated the serf to work on the feudal land part of the time and his own land the rest of the time. The cost to the serf, i.e., working for the lord, was traded against the cost to the lord—providing land and protection. This arrangement was regulated through mechanisms of obligation and punishment (Burawoy 1979). Early capitalism, which freed labor from its bonds to the land, devised new mechanisms to ensure the production of value. Bringing workers under the same roof, it used spatial and temporal techniques, such as clocking in and timesheets, to guarantee the extraction of value from labor. The cost to the worker—partially unpaid labor—was traded against the cost to the capitalist—providing the means of subsistence. This arrangement provided a practical solution to the problem of control, although it incurred new costs (factories, warehouses, supervision, and so on).

The expansion of the scope of the operation of capitalist enterprises transformed them into multi-unit, hierarchical organizations, exacerbating the problem of control into a "crisis of control" (Beniger 1986). The evolution of the railroad system in the United States represents this phenomenon in a very clear manner. As railroad operators expanded their scope from 50 to 100 miles in the early nineteenth century to thousands of miles by the end of the century, running long stretches of railroad networks in different directions, there was an obvious need for coordination in terms of organizational and technological standards, rights of way, freight transportation, scheduling, and the like. Before the expansion, freight moving from Philadelphia to Chicago had to be unloaded and reloaded as many as nine times (Chandler 1980). To respond to this inefficiency, railroad companies created bureaucracies with two or three layers of middle managers running operating divisions, geographical offices, and central headquarters. Chandler (1980) names some of the advantages of this type of organization—economies of speed (cutting costs through standards and routines), flexibility in responding to customer needs, and a steady flow of cash, which reduced the cost of loans and credits. The disadvantage was that the layers of bureaucracy imposed huge expenditures for office buildings, accounting, training, and other overhead costs.

The transition in the later part of the twentieth century to decentralization and flexibility provided a scheme to adjust and reduce these costs. With the service sector and "knowledge organizations" coming to the forefront, and with business and management gurus advocating "liberated firms," Total Quality Management, and project-based units of work (Peters 1992), the best solution to the problem of control was found to be "self-control"—that is, for people to control themselves. This "solution" speaks directly to the growing prominence of notions such as "workforce participation" and "intrinsic motivation," which sought to highlight "the desire to do the work and the pleasure of doing it," as opposed to systems of external rewards (Boltanski and Chiapello 2005, p. 80). The excitement in the United States and the rest of the world around Toyota and its "lean" model of organization largely derives from its successful implementation of notions based on participation and intrinsic motivation (Womack, Jones, and Roos 2007).

For our purposes, these developments find parallel incarnations in a type of heteromation we dub "organizing labor" because it draws on the human capacity to organize activity. Here we understand "organizing" in the broad sense of goal-directed social entities designing and implementing coordinated activity systems linked to the external environment (Daft 2011). We discuss two cases where organizing labor plays a prominent role in the extraction of value.

9.2 Customer Reviews: Transferring the Cost of Control

The implementation of self-control in organizations takes two forms: peer control and customer control. Peer control replaced the direct supervision of the old hierarchies with the evaluation of peers and team members, organized in projects, looking over each other's performances. Customer control extends this model beyond the boundaries of organizations to the appraisals of customers, whose "satisfaction" has been portrayed as the holy grail of business practice for decades. The expansion of the internet and e-commerce has brought customer control into prominence, to the point where one can hardly interact with companies online or offline nowadays without being asked to provide feedback on one's "experience."[1] The rhetoric of customer satisfaction and experience, while partly genuine, hides the more important benefits of customer reviews to business enterprises.

A survey of business literature on this topic illustrates this quite clearly. The list of the substantial benefits of customer reviews to businesses includes referrals, repeat business, loyalty, retention, customer acquisition, competitive advantage, reputation, sales opportunities, morale and motivation, and education (Letts 2013). Some consider reviews "a great SEO [Search Engine Optimization] boost" for business, arguing that even "negative reviews aren't actually bad; they help establish the authenticity of your product" (Lentejas 2013). What is notable here is that the concern of these commentaries is how businesses stand to benefit from customer reviews, discounting the time and labor that customers put into them, despite the clear economic value of the reviews, as acknowledged by the businesses themselves. There is, needless to say, potential benefit to the customer in terms of the information gained from peer reviews, but such benefits pale in comparison to the profit-seeking benefits for enterprises.

9.3 Citizen Science: Harnessing the Crowds

We turn to an analysis of organizing labor in citizen science, an arena very different from commerce in its goals, norms, and outcomes. A relatively long tradition of "community archaeology," for instance, has engaged citizen volunteers in various kinds of projects, some of which are run by amateurs (Simpson and Williams 2008). The Edgware Junior School in London, for example, which was the site of an air-raid shelter in World War II, became an archeological site in 2006. Prompted by the accidental discovery of a concrete structure in the school's playing field, a team of excavators joined the local archeological society in a project with the pedagogical aim of teaching students about the history of the period, as well as introducing the principles and methods of excavation and archeology more broadly (Moshenska 2007).

Largely successful in its aims, this project is an example of an organization "with little or no professional involvement and guidance" (Simpson and Williams 2008, p. 74). As the authors point out, however, the general idea of "archaeology by the people for the people … is something of a naïve fantasy [because] in reality, community archaeology is censored and manipulated, and communication of information and access to the past is controlled through many different agencies" (ibid., p. 72). This situation begins to sound like the corporate governance of the modding community, where mods can be squelched instantly if a company does not like them.

The introduction of Web 2.0 technologies into archeology seems to have exacerbated these issues. Based on a survey of a broad range of archeological projects that leverage social media for public participation, Perry and Beale (2015, p. 154) found that

> Discussions of power, labour, consumption, and capitalistic control that fuel much archaeological web-based engagement are often nonexistent or undertheorized by developers, users, and other implicated parties.

Drawing on the notion of heteromation, Perry and Beale (2015, p. 159) note that "via the social web, volunteers are regularly drawn into the project of resolving profound internal organisational problems: bureaucratic, personal and workflow oriented." They criticize the "exploitative politics" behind citizen projects that do not provide actual learning experiences, "instead using amateurs and beginners as 'workhorses'" (p. 158).

One such project, for instance, makes the following recommendation: "Crowdsourcing application must be *designed carefully to give the impression* that participants volunteer and are not working for free" (p. 158; emphasis added by Perry and Beale). The "exploitative politics" behind such statements, according to these authors, is hidden behind the volunteers' desire for personal fulfillment and socialization—an antidote to the professional labor bottleneck that has arisen in the era of austerity in Britain and elsewhere. In this light, Perry and Beale conclude that the gains to disciplinary and institutional centers are much more discernible than benefits to individual volunteers.

Other studies of citizen science have shown a smaller number of projects that enable fuller forms of participation, higher levels of engagement, and a balanced distribution of rewards compared to the greater number that disallow or limit contribution and recognition (Cornwall 2008). Qaurooni et al. (2016) studied a wide array of citizen science projects, discovering a similar bias toward what they call "Crowd Science" compared to the more collegial and participatory "Civic Science." As examples of these types of citizen science, the authors discuss two map-making projects, Cropland Capture[2] and Kite Mapping,[3] both of which draw on citizen help for map construction. The projects differ on key dimensions such as citizen roles, organization, and technology. Cropland Capture recruits citizens to remedy scientists' poor understanding of cropland locations across the globe, having them evaluate scientist-selected land areas through a simple yes-no question, using a gamified smartphone application and geotagged pictures. Kite Mapping, on the other hand, considers citizen participation as a means to contest official maps and challenge their beneficiaries. To that end, it allows people to pick the time and place of their interest, provides them with guidelines and support for buying, assembling, and flying a map-making kite using simple household items, and equips them with tailor-made software to stitch together their own map of an area. The result is then submitted to the public record and may be used for advocacy and activism, as well as research (ibid.).

The contributions of volunteers in a significant number of citizen science projects include activities such as organizing or hosting groups, classes, and workshops for topics such as climate change, environmental monitoring, volunteer training, brainstorming sessions, solving math problems, raising funds, and collecting debris. Organizers and participants of these events often report their findings to a lab or institution without having access to the bigger picture of how their findings are incorporated into scientific work and scholarly publications. Study after study has shown that there are significantly more citizen science projects that are like Cropland Capture than Kite Mapping (ibid.). The distribution of the projects illustrates a pattern that is observed in increasingly different sites and areas, as we see next.

9.4 The Invisible Participant: From the Bell Curve to the Power Law

Customer reviews and citizen science provide two examples of a larger trend in a constellation of activities referred to as "crowdsourcing," "wikinomics," or "human computation." A common resource of these activities is the self-organizing capacity of human beings, which provides an alternative model to old hierarchies, eliminating or minimizing the cost of control and coordination (Benkler 2002). The trend reflects the potential of Web 2.0 technologies for facilitating coordinated activity at minimal cost. Shirky (2010) provides a large set of examples, including Wikipedia and its concept of collaborative content creation; the Linux operating system and its open-source approach to programming; the Voice of the Faithful campaign against child abuse in the Catholic church; the Coalition for an Airline Passenger Bill of Rights; the successful campaign of the student customers of the HSBC Bank to revert an overdraft penalty; Howard Dean's supporters in the US presidential campaign of 2004, who did some of the first online political organizing; and the use of Twitter by political activists in Egypt.

The list is, indeed, compelling in giving us a glimpse of the power of what Shirky calls "social tools" in enabling new kinds of "organizing without organizations." The power derives from the capacity these tools provide for coordinated action across time and space, at large scale and high speed, and with almost no cost. In spite of these advantages, however, the majority of participatory projects do not succeed, failing to accomplish their goals or sustain themselves. In fact, the performance of projects seems to reveal a pattern similar to what happens in citizen science. This pattern can often be expressed in the form of a power law distribution—that is, a distribution of projects with "a lot of failure, some modest success, and a few extremely popular [projects]" (Shirky 2010,

p. 235). Of about 100,000 open source projects on SourceForge.net (a site that hosts many such projects), for instance, very few are downloaded millions of times, and the majority don't get even a thousand downloads. Below the 75th percentile, projects have no downloads whatsoever (ibid., p. 244). Shirky explains this pattern in terms of the low cost of failure which allows projects to explore and experiment with different possibilities, essentially "for free." Terms such as "success," "failure," and "free," are relative descriptors, however, leading one to ask, Free for whom? and, Who bears the costs of failure or the benefits of success?

To answer these questions, we need to consider another power law distribution that is as commonly observed *within* these projects as it is *among* them. This distribution has to do with the division of labor among participants. Numerous studies have shown that the bulk of the work in participatory projects is done by a very small percentage of participants, with the majority contributing little. This is the case whether we consider Linux programmers, Wikipedia authors and editors, or people who tag photos on Flickr (Shirky 2010, p. 124). Shirky explains this "predictable imbalance" in terms of differences in motivation: "The number of people who are willing to start something is smaller, much smaller, than the number of people who are willing to contribute once someone else starts something" (ibid., p. 239). Motivational differences, according to this line of thinking, lead to a "spontaneous division of labor," where "no effort is made to even out ... contributions" (ibid., p. 125). Actually, however, the bigger point is that this spontaneous division of labor "wouldn't be possible if there were concern for reducing inequality ... [because] most large social experiments are engines for harnessing inequality rather than limiting it" (ibid.). Not only should this "natural" state of affairs in terms of motivational differences and inequality *not* be discouraged, it should be maintained and encouraged because it guarantees the emergence of a small number of "a few extremely popular" projects. Inequality is endorsed as natural and inevitable.

We find the sense of generality, spontaneity, and inevitability emanating from this logic problematic. First, to use individual motivation as the explanation of these phenomena inverts the causal order, putting the (social) horse of predicaments and possibilities behind the cart of individual psychology. It is true that people vary in terms of their motivations, but the generators and enablers of those motivations can be largely found in the socioeconomic circumstances of people's upbringing and current life circumstances. Rather than attributing differences in participation level to differences in "intrinsic" motivations among individuals, therefore, it makes more sense to think of this in terms of their ability to maintain commitment to particular goals. If we understand committed behavior as the capacity to engage in "consistent lines of activities" (Becker 1960), then varying behaviors can be explained according to the outcomes they produce and the system of values that is applied in assessing those outcomes: "What kinds of things are conventionally wanted, what losses feared? What are the good things of life whose continued enjoyment can be staked on continuing to follow a consistent line of action?" (ibid., p. 39). Psychologists and behavioral economists have documented, for instance, the effect of poverty on mental stress, with deep influences on people's willingness to set goals and on their commitment to work hard to accomplish them (Haushofer and Fehr 2014). Rather than motivation being an inherent attribute of individual psychologies, it is in large part the outcome of socioeconomic circumstances. For example, Körner, Reitzle, and Silbereisen (2012, p. 190) point out that, "Even in wealthy, stable countries with a strong safety net such as Germany, unemployment may lead to a downward spiral of psychological distress."

Second, the "spontaneity" of the division of labor is relative to how group relationships are configured, resources are divided, and information is distributed. "Wikipedia, which looks like a reference work to the average viewer, is in fact a bureaucracy given over to arguing," Shirky comments (2008, p. 278). Wikipedia has around a dozen administrative collections of pages, only one of which is for the actual articles (ibid., p. 279). What seems on the surface to be a spontaneous arrangement is actually a structure implemented to maintain control of the editing process and the integrity of the content. Forte, Larco, and Bruckman (2009) found that the Wikipedia community governed itself by articulating its norms as rules and strictly enforcing those rules (ibid.). This structure, embedded into the Wikipedia technology, along with resistance toward commercialization (and the financial backing of Jimmy Wales), have enabled Wikipedia to survive as an open system for many years. Wikipedia is not the result of "motivated individuals," but of commitments that give rise to strong organizational and disciplinary structures.

While Wikipedia has thus far adhered to its founding principles of collaborative and open content creation, not all systems are founded on such principles.[4] And this brings us to asymmetries in the distribution of rewards that is becoming increasingly prevalent in participatory projects. The asymmetries derive from a seeming paradox in the relationship between individuals and groups.

Describing the most highly connected people as the backbone of social networks, Shirky (ibid., p. 213–214) argues that these networks "are held together not by the bulk of people with hundreds of connections, but by the few people with tens of thousands."[5] A few pages later, he acknowledges: "Perhaps the most significant effect of our new tools, though, lies in the increased leverage they give *the most connected people*" (ibid., p. 225; emphasis added). These statements speak to a paradox: Is the powerful position of the highly connected people an inevitable outcome of differential motivations among people, or is it an artifact of the design of current networks and their "winner-take-all" incentive structure?

In the absence of empirical data based on a thorough exploration of both scenarios, we might not be able to settle this question. We do know, however, which alternative is more compatible with the spirit of capitalism. In particular, developments in the last few years show a clear trend toward the appropriation of participatory projects in favor of the interests of a select few at the expense of the large majority. The wealth accumulated in the last decade by technology companies such as Facebook, Google, Amazon, and others illustrates the "resolving" of the paradox toward the design of systems that have turned a very small percentage of people into the major beneficiaries of these projects. This trend provides an interesting example of how capitalist institutions have by and large managed to co-opt the outcomes and mechanisms of participatory projects in various activities. In so doing, capitalism has, yet again, turned a predicament into an opportunity for its own gain. Therein lies the true nature of the power law distribution: the long tail gets the short shrift.

A telling example comes from the case of the Canadian firm Goldcorp Inc. Around 2000, when gold markets were shrinking, Goldcorp's 50-year-old mines on a property in Red Lake, Ontario, were depleted of cheaply extractable deposits, putting the company on the verge of bankruptcy. Having learned about Linux, the new CEO devised a plan for opening up the company's databases to allow a participatory competition to identify promising locations for excavation. With a total purse of $575,000, an army of geologists, mathematicians, students, and other folk identified 110 targets on the 55,000-acre property, 50 percent of which had not been identified beforehand. The outcome of this initiative was turning a $100 million company into a $9 billion giant! Prize money of a few hundred thousand dollars, in other words, gave rise to many billions of dollars. Considered a gold standard, if you will, of participatory projects, the case of Goldcorp Inc. reveals the dynamics of relative rewards in power law distributions and how they hugely favor capital.

In this light, the increasing prominence of power law distributions, which has apparently come at the expense of bell curves and their normal distributions, should be understood as an outcome of expanding computer-mediated networks and the kinds of social arrangements they support. Power law distributions, in other words, serve the same analytic *and* normative function in networked environments that bell curves provided in hierarchically layered environments. They normalize a radically non-symmetric distribution of rewards, wealth, and prestige while bell curves normalized the "average" and the "normal." In an ironic twist, however, while the "normal" distribution of bell curves hid individual idiosyncrasies, power law distributions do the opposite, accentuating individual differences and aggravating inequality. While old-fashioned statistics could "lie" by burying individual differences under the shadow of averages (Huff 1954), newfangled Bayesian statistics have the potential to bury a large percentage of the population under the logic of networks, rendering their contributions to the working of networks invisible. The old economists' joke—that Bill Gates walks into a bar, and the average income of everyone becomes $1 million—loses its punch, giving way to a more current version: Mark Zuckerberg walks into a bar, and everyone likes everyone, but no one has enough money to pay for their drink. Then Eric Schmidt walks in and invites everyone to search on their Androids for moneymaking ideas using Google's real-time analytics. Finally, Jeff Bezos walks in, and offers everyone a HIT!

9.5 The Paradox of Participation: New Predicaments and Possibilities

The questions and problems introduced by organizing labor embedded in a capitalist economy highlight the ambivalent character of computing technology in a rather vivid manner, facing us with what can be described as the paradox of participation: Should we, or shouldn't we, as individuals and communities, partake in these projects? Refraining from participation would deprive us of an effective mechanism of innovation, contribution, and community building. Taking part, on the other hand, not only might enhance social and economic inequality, it can expose us to unpredictable risks and consequences. While a sizable portion of the population seems to have resolved this dilemma in practice, favoring participation and contribution, the situation is more complex. The complexity is evident in a new initiative called Human Computation, which seeks to leverage the potential of computer-mediated participation by structuring and formalizing the division of labor between humans and machines in an explicit way, and arguing for its pervasive implementation throughout the economy and society. The *Handbook of Human Computation* (Michelucci 2013, p. 85) defines the field as one that considers "the design and analysis of multi-agent information processing systems in which humans participate as computational elements." The idea of humans as "computational elements" provides a tellingly accurate depiction of what is involved in organizing labor and the mechanisms that make human labor invisible. While sounding innocuous, this conception of the place of human beings in the big scheme of things runs the risk, especially in a capitalist system, of undermining basic human rights and values. If industrial capitalism turned individuals into little cogs in huge manufacturing plants, giving rise to the predicaments of alienation and isolation, current capitalism turns people into "bits" of information, pieces of wetware, or simply abled bodies ultimately in the service of others, usually powerful others. The cycle of innovation and co-optation repeats itself, and generates new possibilities and predicaments.

What this implies for the future for the future of human individuals, communities, and societies is a question that we pick up in part III of the book.

III
Looking Ahead

10
Mechanisms of Participation: A Story of Rewards (and Punishments)

10.1 Introduction

The varieties of heteromated labor examined in part II exemplify a broader trend in capitalist economies toward less visible, more loosely structured but digitally mediated means of extraction of value. These are part of the response of capitalist societies to the predicaments we have discussed. The response, which often takes the form of co-optations of earlier possibilities, generates, in turn, new predicaments and possibilities, repeating the cycle of capitalist change, as discussed in chapter 3. We want to examine these new predicaments and possibilities in this last part of the book, analyzing specific mechanisms of reward and punishment that emerge from heteromated labor that entice individuals to continue to participate, or, in some cases, force them to. These mechanisms largely constitute the light and positive side of heteromation, showing us the possibilities presented by this cycle of capitalist development, but they also include delicate mechanisms of control and punishment. In simple terms, it is the perception of these mixed possibilities that induces individuals to participate in heteromated labor. We focus here on the question: *Why do people participate?*

To answer this question, we cross-tabulate the varieties of heteromated labor and activity against the rewards and punishments that accrue to each. We have compiled an opportunistic sample in table 10.1, based on the cases examined in earlier chapters (and a few others), to demonstrate the broad and diverse range of activities and mechanisms.

The mechanisms comprise three logics of activity: convenience, care of the self, and social relations.[1] The logics are not completely mutually exclusive, but we use them as organizing concepts to show trends and patterns within the mechanisms.

Table 10.1

Mechanisms of Heteromated Labor by Subject Type

● Bubble stars   ○ Less visible participants
underscore designates variable influence of the mechanism

SUBJECT TYPE	remuneration	convenience	totalized stimulation	social connection
Mechanical Turkers	○			
Correspondent Bankers	○			○
PARO Caregivers	○			<u>○</u>
Phone Menu Users				
ATMs & Self-Service Users		○		
Google Searchers	○	○		
Goldcorp Contest Losers				
Design Contest Losers				
YouTube Followers			○	○
YouTube Non-Celebrities			○	○
Citizen Scientists		○	○	○
Health I.Q. Participants		○		
League of Legends Judges				
Political Essayists			○	○
Recommenders & Reviewers			○	
Video Game Modders			○	○
Social Media Participants		○●	○●	○●
Video Gamers			○●	○●
Design Contest Winners	○●			
Goldcorp Contest Winners	●			
YouTube Celebrity Creators	●			●
Spacewar! Developers				●

162 Chapter 10

Mechanisms of Participation

community	altruism	self-expression	micro-validations	self-improvement	coercion
				○	
○	○				
	<u>○</u>				
					○
					<u>○</u>
				○	
		○		○	○
		○	○		
	○	○	○	○	
○	○	○	○	○	
		○	○	○	
○	○	○		○	
○		○	○	○	
	○	○	○		
○		○	○	○	
○<u>●</u>		○●	○●	○●	
○●	○●	○●	○●	○●	
		○●	○●	○●	
		●	●		
		●	●	●	
		●		●	

10.2 Convenience through Access

The most obvious and least complicated logic is that of convenience. Computer-mediated information search and communication are undeniably convenient, and so ubiquitously threaded throughout the daily round that our reliance on them is all but complete. Participants seeking convenience utilize applications such as search and social media that draw in billions of users.

Self-service is, relatively, more of a mixed bag: technologies such as ATMs afford considerable convenience, but some devices and applications render tasks more difficult than they would be with human assistance, forcing participants to traverse predefined digital programs of interaction to reach a goal for which there is no other path to completion. These programs are designed according to a binary logic with a narrow set of options that suits computers: If X, then A, else B. As a result, more often than not they channel individuals into actions not of their own choosing. If one is just ordering a hamburger at a kiosk, this is fine. In the case of, for example, phone menus needed to access critical services, repeated attempts (if needed) must be passed through anew from the beginning, tediously following the entire path. Because many services are accessible only through phone menus, they can be quite baffling and frustrating. Costs are pushed onto service seekers (such as customers or patients) who pay the price of their time, and who may find the process wearying or ineffectual in solving their problems, leading to further costs, such as not settling a bill or receiving a needed medical service. In these cases, the organization deploys its power to impose costs (which are cost savings for them).

Though the "punishments" of heteromated labor are not as common as the rewards in table 10.1, loss of time, money, and autonomy, as well as breakdowns in task completion, occur fairly frequently, sometimes with serious consequences. These phenomena differ from the punishments of failure to comply with disciplinary regimes that Foucault spoke of. Such punishments, sinister in their willingness to inflict precisely crafted forms of torment, are, nonetheless, comprehensible, transparently making known their intentions. By contrast, the dispersed, diffuse, hidden intrusions on our resources and sense of well-being from the coercions of digital technologies, render them less subject to critique, more difficult to apprehend and assess, and perhaps contributory to a vague sense of alienation and confusion. The intrusions often seem meager, yet they reveal institutions that do not care about us.

However, most of the mechanisms of table 10.1 draw us into heteromated labor with small but pleasing rewards, relying on opportunities that we choose to accept that permit us to get through our days more easily, and often to feel better about ourselves. These mechanisms speak to the logics of creating and maintaining social relationships, and to care of the self.

10.3 Care of the Self through Technologies of the Self

Key logic of participation can be broadly understood as care of the self. Foucault's notion of care of the self, developed in his later work, establishes this care as a "way of living" or an "art of existence" (1986, p. 43). Care of the self calls for tending the "soul," or more particularly, "the activity of the soul" (1997, p. 230). Foucault spoke of individuals seeking "a certain state of happiness, purity, wisdom, perfection, or immortality" (1997, p. 225). In his work on care of the self, Foucault relaxed his emphasis on coercive disciplinary mechanisms. He began to consider a person's role in transforming himself or herself, while noting that this transformation still occurs within certain regimes of power (ibid.). The "certain state of happiness ..." is cultivated through "technologies of the self" that "permit individuals to effect, by their own means or with the help of others, a certain number of operations on their own bodies and souls, thoughts, conduct, and way of being" (ibid.). The internet can be broadly conceived as one such technology.

Though conventionally we take the internet to be a gigantic global machine of social connection (which it undoubtedly is), a close reading of table 10.1 suggests another, more shadowy figure moving within—that of the self, surfacing in active engagement with the mechanisms of totalized stimulation, self-expression, self-improvement, and microvalidation. Starting at row 8 of the table, self-expression figures in every type of heteromated labor we examined, appearing in multiple and varied forms. Microvalidations are also abundant—apparently the tiny moments in which we are noticed online mean a great deal to us, and we set great store by the bestowal of favorites, retweets, recs, likes, upvotes, comments, replies, karma points, +1s, and other clickable means of demonstrating approval and interest. Self-improvement entails deliberate actions undertaken to enhance some personal quality—for example, to become more knowledgeable (citizen science), more community-minded (Tribunal participation), or to obtain a healthier body (Health IQ activities). Some Mechanical Turk workers report that their work affords self-improvement as a sort of by-product of making money. For example, they may learn about research by filling out surveys, improve their skill in a second language (in the case of workers from India), or become more aware of current events through the HITs they complete (Jiang, Wagner, and Nardi 2015). The mechanisms of totalized stimulation, self-expression, self-improvement, and microvalidation signify desires to assert, construct, project, develop, and nurture a self.

10.3.1 Stimulation through Entertainment

Totalized stimulation offers the most obvious program of care of the self—we use digital technology to alleviate and displace boredom, anxiety, fear, confusion, lack of ease. We seek entertainment and pursue restorative hobbies in actions that constitute the "operations on our bodies and souls" through which we aim for happiness.

However, it is also the case that such activities may "overwhelm" us, to use Dewey's term, when they become runaway passions, transmogrifying something good into something much less good. Such mutations of activity seem an unavoidable result of living. We face hazardous existential conditions, implicit in the specific predicaments we describe, and, more broadly, in the human propensity to find difficulty balancing competing aims and impulses. The culture may rework these issues of balance, however imperfectly that happens, through periods of reflection.

Totalized stimulation, as such, is ambiguous in its impacts and outcomes. Access to quality entertainment is a beloved feature of the internet, and drives heteromation across virtually every demographic. But such access has also given rise to memes of addiction and obsession. The idea of "internet addiction" began in 1995 when psychiatrist Ivan Goldberg spoofed the culture's narcissism of endlessly searching for things that might be wrong with us. Despite Goldberg's satiric intent, the idea of internet addiction struck a nerve and has persisted. It is invoked with respect to the easy availability of pornography, and recycled periodically as a putative driver of violence (in the manner of recurring "moral panics" documented by sociologist Stanley Cohen (1973)). For example, in July 2011, when Anders Breivik shot 77 people in Norway, the addicted gamer stereotype cropped up (on cue). A Reuters news story reported that Breivik's attorney identified Breivik as a "loner" who "played video games" and was "probably mad."[2] Coop Norway, one of Norway's largest retailers, removed 51 video games and weapon-like toys from its shelves in reaction to the incident.

Programs, camps, and centers for internet and video gaming addiction exist, in varied forms, in the United States, Europe, China, and Korea. For a brief period, electroconvulsive (shock) therapy was deployed in a clinic in China to "treat" young people supposedly spending too much time playing games. This coercive regimen and its sociopolitical implications were immortalized in a satiric agitprop machinima produced by the Oil Tiger Machinima Team, a media-savvy group of Chinese World of Warcraft players. In the video, the villainous doctor Yang Yongxin administers shock therapy to youthful video gaming "addicts." The machinima was a somewhat surprising expression of digitally based resistance in a country with tight controls on critical commentary. For a period after its release, the video was watched by millions of Chinese, not just gamers, and then again following a ban the government quickly rescinded.

The long-term effects and implications of totalized stimulation are unknown, and while we doubt that in most cases they constitute true clinical addiction, the meme reveals persistent anxieties about digital technology. Like the Chinese authorities, our wider culture suspects that there may be something very wrong with certain forms of internet usage. Notions of internet addiction will probably continue to influence popular and scholarly interpretations of the impacts of digital technology (see Hellman et al. 2013). But we must also note its relation to the current ideal of a self that is always stimulated, engaged, and driven. The notion of addiction bears a cautionary tale of the self overshooting attempts at self-assertion and self-nurturance. Foucault himself, for all his self-awareness, was, in the end, overwhelmed. Totalized stimulation offers its rewards with a sharp edge—and that is part of its appeal, and part of its danger.

10.3.2 Self-Expression through Reporting

Technologies of the self involve tools and procedures we deploy in efforts to come to know ourselves. A large part of our identities is shaped by our capability of providing life narratives—a capability that has been largely undermined in the precarious and transitory lives that contemporary humans live (Sennett 2007). Digital technology has stepped in to fill this gap. Blogging, for example, allows participants to "document their lives"; to express opinions in order to influence others and to receive others' opinions and feedback; to practice "thinking by writing"; to post about emotions to "work out [their] own issues"; and to provide commentaries on topics they care about (Nardi, Schiano, and Gumbrecht 2004, pp. 222–225). Murthy (2012) suggests that the act of regularly tweeting declares the existence of oneself within digital space. Foucault pointed to the daily records we produce that detail not only major life events, but also, and more crucially, the everyday and the trivial (1997). Murthy (2012, p. 1070) discerned this pattern in Twitter usage, reporting: "For example, when people follow the tweets of those they have met at conferences, they will most likely be exposed to their daily music listening habits, sports interests, current location, and shopping wish-lists, amongst other things."

Such trivial reports are, notably, the stuff of everyday, face-to-face interaction, shifted to the digital sphere. Often, these reports become particularly meaningful when we find ourselves separated from others with whom we would have shared, in person, the series of these small moments that make up a life. With digital mediation, there is also an extension, in more novel fashion, to those with whom we would *not* share such moments (such as someone we met at a conference), altering the character of everyday interaction, but at the same time, affording new opportunities for social relationships.

10.3.3 Self-Validation through Micro-Mechanisms

Acts of self-expression ("effected by our own means") and incorporation ("into our souls and thoughts") of the responses transmitted by the microvalidations ("with the help of others") contribute to the project of knowing the self. While this undertaking is a social process, it is one that carves off the individual self as the singular focus of activity, separating the individual from others through assertion of a *particular* self—particular opinions, commentaries, preferences, emotions. Identity is formed and reformed as we express ourselves and absorb the microvalidations, permitting us to distinguish ourselves from others. If we instead desired a feeling of unity, a seamless merging with equal others, we would not engage in constant acts of self-expression or seek the affirmations of the microvalidations. Heteromated labor thus reflects us back to ourselves in ways that help us understand who we are—an important task in the midst of the muddles and complexities of the predicaments. Crystalline, definitive moments in which we are able to express ourselves, or to experience the satisfactions of the microvalidations, counter the instabilities characteristic of the predicaments.

Tellingly for us in the contemporary moment, Foucault did not speak of security in suggesting how to cultivate care of the self. He put forth a different set of intentions and aspirations—to seek a state in which to be pure, wise, perfect, even immortal! Could it be that the lofty aims of the care of the self offer a robust response to the dislocations of the predicaments? Could these aims be a counterpoint to the reality in which most of us dwell most of the time within the anonymous vastness of the long tail? A Chinese writer said, simply, "If there were no internet, it would never have been possible for me to have my words heard" (see Zhou 2005, p. 779). Most of us share a similar sense of wonder and gratitude at the technology we have chosen to weave within our lives. It allows us to project our selves out, and grounds us in a more complex reality with at least some moments that nurture and sustain us.

10.4 The Power Law Society and Its Winners

The power law distribution, as we saw, involves a small number of prominent individuals within its tall head, trailed by a large number of individuals within its long tail. In social situations where the power law distribution of wealth, reward, or recognition occurs, very few individuals reap the benefits, at the expense of everybody else. This reality is often obscured by the (meager) rewards accorded to those in the long tail, such that the extremity of the imbalance can be ignored as one focuses on the rewards.

The power law distribution has prompted many observers to comment on the implications of this asymmetric and often absurd distribution of benefits. Brynjolfsson and McAfee (2016) take up this theme, discussing what they call the current "winner-take-all" economy. They memorably demonstrate the absurdity the power law distribution may lead to:

> A gold medal winner at the Olympics can earn millions of dollars in endorsements, while the silver medal winner—let alone the person who placed tenth or thirtieth—is quickly forgotten, even if the difference is measured in tenths of a second and could have resulted from a gust of wind or a lucky bounce of the ball. (2014, p. 151)

Brynjolfsson and McAfee (2016) observe the power law distribution for the wealthiest individuals who make up the tall head of the distribution, such as sports and media stars, corporate owners and managers, bankers, and lawyers. Within this group exists an even more elite "1 percent of the 1 percent," as Brynjolfsson and McAfee call it (2016)— those at the pinnacle of our socioeconomic system who have garnered historically unprecedented amounts of wealth, power, and recognition.

10.4.1 The Power Law and Heteromation

Brynjolfsson and McAfee (2011, 2016) argue that automation is causing conventional jobs to disappear. This appears to be true, and the authors construct a careful case for this historic change. But job loss is not the same as *labor* disappearing. Heteromated labor is a big part of what allows wealth to flow to the 1 percent, and to the 1 percent of the 1 percent. The authors have overlooked this labor, pointing to automation and high-level social networking as the prime enablers of the accumulation of wealth. Brynjolfsson and McAfee say, for example:

> [T]he fifteen people who created Instagram didn't need a lot of unskilled human helpers and did leverage some valuable physical capital. But most of all, they benefitted from their talent, timing, and ties to the right people.

However, from the point of view of heteromation, we would say that the fifteen people *absolutely did benefit*, and continue to benefit, from an immense reservoir of labor ("unskilled human helpers," in the authors' terms)—without which Instagram could not possibly function.[3] That such heteromated labor is unpaid does not mean that it does not exist. It does exist, and it contributes to the bottom line. One of the key points in this book is that labor is changing—it is not fading away. Labor continues to provide value to capital as it always has. Personal social networks and digital technology are not the only forces that allow the 1 percent of the 1 percent, such as Eric Schmidt and Mark Zuckerberg, to accumulate amounts of money so vast and distributed we cannot even comprehend, much less reasonably tax, them. The presence of heteromated labor does not contradict Brynjolfsson and McAfee's point that conventional jobs are disappearing; rather, it generates a new set of questions and possibilities to consider regarding fairness, employment, and how to organize and manage society.

Schmidt, Zuckerberg, and a few others of their ilk are the big winners of contemporary capitalism, deriving their wealth from the heteromated labor of billions of individuals around the globe. The 1 percent of the 1 percent, whom Brynjolfsson and McAfee also refer to as "superstars," includes a diverse set of individuals: founders and CEOs of tech startups such as Instagram and Intuit; J. K. Rowling, the author of *Harry Potter* and the first billionaire author in history; CEOs of major corporations. The mechanisms producing these superstars are digitization and globalization—digitization because it reduces the cost of accessing intended audiences (customers, readers, employees) to almost nil, and globalization because it has broken down geographic barriers.

Heteromation has given rise to another, less dazzling, but still important group of "winners" who also benefit from others' heteromated labor, through a somewhat different mechanism. The distinctive mechanism is rather novel and intricate, working through the ideal type of contemporary capitalism. We call these people "bubble-stars."

10.4.2 Bubble-Stars and Their Followers

The ideal type of the current era, as we have discussed, is engaged, entrepreneurial, ambitious, focused, competitive, and committed to a vocation or avocation (the successors, it would seem, of the 1980s go-getters we discussed in chapter 1). Within the group of heteromated laborers, those who embody these characteristics constitute a tiny but important and visible minority. They include, for example, winners of the graphic design contests, elite modders who hobnob with game developers, and celebrity YouTube creators. Here, we are concerned strictly with the population of heteromated laborers (not the owners and managers of capital), and the pattern within which a small number of these laborers escape the meagerness of the typical rewards of heteromation.

We identify this character type in, for example, LPers such as PewDiePie, described in chapter 5. Turning heteromation on its head, these few individuals leverage the system for their own gain. They attain recognition and sometimes money (though money does not appear to be the initial motive[4]) through their labor.[5] They eclipse marginalization by drawing on a natural abundance of intelligence, wit, personality, persistence, and luck.[6] Although sparsely represented in heteromated systems, the vitality and success of these participants galvanize masses of others, such that the visible participants themselves become "mechanisms of participation." We see these participants in the lower portion of table 10.1, while those who earn money are to the left.[7]

The sharp power law split ensures that only a tiny minority of participants can succeed in this way—unlike the normal distribution of previous eras of capitalism, in which the largest space under the bell curve was occupied by the "average" person. The visible persons of today function in some ways as the post–World War II stars of Hollywood and the music industry did, rousing people to take part in the new economy. Like movie stars, they exist because the crowds are drawn to their élan and energy. They regenerate and refresh heteromated systems with the continual flow of their ongoing activities. Rewarded with recognition, and sometimes remunerated, they perform for the crowds, producing, for example, desirable software modifications, amusing Let's Play videos, provocative Twitter feeds. The long tail, by virtue of its sheer numbers, drives the economic value, and the visible participants drive the masses.

There are, however, key differences between these new, visible, energetic participants and the stars of the past. First, the new stars operate within "internet bubbles"—the echo chambers that unite people with similar cultural tastes, ideologies, or identity politics (see Jamieson and Cappella 2008). We can think of these new participants as "bubble-stars"—sensational persons operating within bubbles. Unlike the era of massification, in which the masses watched the same television shows and followed the work and lives of the same movie stars, baseball players, and so on, the divisions that broke apart the unified crowd in the 1980s led individuals to follow only certain performers and celebrities. While virtually everyone knew who Elvis was, not everyone knows who PewDiePie is. His subscribers comprise a defined demographic, and he is unknown outside of it. This development, in which bubble-stars are bound to specific types of others, is consistent with the neoliberal doctrine of the heroic individual with the freedom to choose. We can choose which performers we prefer, which fan fiction to write, which games to play, which bloggers to read, which YouTube creators to subscribe to. The small number of broadcast channels with constrained, lowest-common-denominator programming that reigned during the era of massification was disrupted by the affordances of computer-mediated networks supporting a plenitude of minutely defined linkages.

Second, in the digital context, bubble-stars are not kept out of circuits of fame by professional social networks, movie studios, credentialing institutions, or the other gatekeepers who otherwise control access to coveted activities. What this "democratization" of culture implies for the status of bubble-stars is that they rely for their vitality on a "following," (such as subscribers) or others being aware of them, (such as peers being aware of contest winners), or those deployings their productions, (such as gamers using mods). The persistent need to capture the attention of others keeps the bubble-stars on their toes, providing incentives to continue delivering value to the commercial enterprises that are behind the scenes, even though the bubble-stars do not work for the enterprises, and even though they think of their followers/viewers/users/ as their own.

Third, a small number of earlier stars, having passed muster with Hollywood gatekeepers, made it into the rarified and permanent world of filmdom, enjoying an almost mythical status—once a star, always a star. Their status gave them not only fame and money, but cultural and political power, if they chose to pursue it. Charlton Heston, for example, was able to drive a specific political agenda as the influential spokesman of the National Rifle Association , an extremely powerful organization in American politics. Bubble-stars are unlikely to hold such sway because of the sociodemographic limitations of the individual bubble within which each operates, and because they can quickly flame out (unlike someone who played Moses and Ben-Hur during the era of massification). Bubble-stars may see their stars rise (and set) early.

10.5 The Masses Unbound

We have said much about bubble-stars and their role in heteromation, yet we must now emphasize that a great deal of heteromated labor—perhaps the bulk of it—occurs within the quotidian pursuits of search, email, writing Yelp reviews, posting to Facebook, online shopping, completing HITs in Mechanical Turk, projects such as citizen science. Here we have no bubble-stars, just people using the internet for everyday tasks and projects. These crowds incite participation too—not in the flashy manner of the bubble-stars—but quietly and modestly, as they industriously contribute to nurturing communities, helping others, and creating social connections.

This labor is concentrated in the top portion of table 10.1. For example, PARO caregivers work at the margins of the robots, never becoming stars. Citizen science is not organized around winners and celebrities, and individual AMT workers may become visible on worker websites or in social media feeds. But they cannot "win" within the activity of microwork.[8] Correspondent bankers' essential activities remain invisible to developers and bank managers. Self-service is nearly always an invisible contribution to an enterprise's bottom line.

These forms of heteromation involve pretty much anyone who uses the internet.[9] Such participation thus gives rise to literally billions of moments of labor—small, transient, unnoticed, yet accruing value through sheer volume, sufficient to sustain gargantuan enterprises such as Google, Facebook, and Amazon.com. Lesser enterprises are sustained too, some of which are eventually subsumed by the giants (e.g., Google acquired YouTube and Amazon acquired Twitch.tv), and some of which do very well on their own.

Thanks to us, that is, all of us who comprise the masses providing heteromated labor, new forms of heteromation seem to continually materialize. Many types of activity support these processes, with "big data" being a particularly versatile and pervasive driver of value. With our dividuated, incessant practices of gazing at ourselves every which way, stimulated by the predicaments, we have prodigious amounts of information to offer. New companies are constantly forming to harvest this information. For example, Health IQ (healthiq.com), a startup in Mountain View, California, offers free daily online health quizzes to improve "health consciousness."[10] The quizzes are not comprised of the stodgy questions that characterize formal education; rather, with considerable flair, the Health IQ writers tap into the quirky richness of popular culture. For example, participants answer a question about diet where the choices are vegetarian, vegan, paleo, Mediterranean, low-carb, Ornish, DASH, calorie-restriction, intermittent fasting, gluten intolerance, lactose intolerance, take supplements. Once the participant's basic information is in the system (and, at this point in the book, we need not point out its value to insurance companies, with whom, predictably, Health IQ partners), the participant can move on to take the other quizzes, which change daily. This flow of activity stimulates a process of continuous production of valuable, up-to-the-minute data.

The quizzes are informative and research-based, speaking forthrightly to real questions concerning how we care for ourselves. Yet they also include many questions regarding very personal matters. There is no mention on the website of how participants' data are stored, assessed, aggregated, analyzed, transferred, shared, or sold.

We might ask how it is possible to come up with a new quiz about the topic of health every day. One way is heteromated labor. Many of the quizzes are tagged with a "Suggested by" graphic bearing the name of the participant who suggested the quiz (for example, "Tea: Big Benefits from One Small Cup," Suggested by Melanie Sanders; and "Fitness: Strength Training and Pain Relief as Men Age," Suggested by Harold Brown). The quizzes themselves are written by paid writers, but Health IQ taps into the crowds who visit their site to generate ideas for the quizzes. Note also the construction of the microvalidation for a correct answer—the font size for "You're right!" is considerably larger than the health-related content below (figure 10.1).

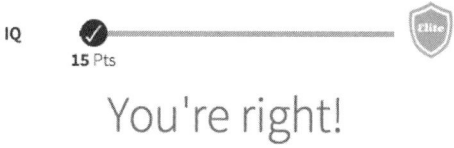

Figure 10.1

The answer to a Health IQ quiz question

This small startup thus leverages heteromated labor in two ways: by stimulating participants to produce monetizable personal health data, and by engaging participants in shaping the design of the quizzes. Through a series of microvalidations (there are leaderboards and other mechanisms too), as well as information useful for care of the self, participants are enticed to provide personal health information.

10.6 The Largest Mass of Workers Possible

In sum, two distinct forms of participation organize people within processes of heteromation: (1) the bubble-stars and their followers/viewer/users, and, (2) individuals leveraging information and applications on the internet in pursuit of everyday projects and interests. Heteromation's adaptable profile fits well within capitalism's strategy of inclusion for purposes of "exploiting the largest mass of workers possible," as Caffentzis (2013, p. 72) put it. The mechanisms themselves are versatile, ranging from myriad "affective" rewards Terranova (2000) described in her early influential work, to more coercive means of extracting value.

Dickens spoke of the best of times and the worst of times—it appears that we might be in such times right now. Deleuze, as we have noted, observed the co-existence of "liberating and enslaving forces." Does this Dickensian, Deleuzian ambiguity mean we should just relax and focus on the positive? Probably not, in our view. We discuss pertinent issues in chapters 11 and 12, and in the epilogue. For now we will allow ourselves to be inspired by much of the activity of the "mass of workers" in table 10.1: the diligence of the citizen scientists, the creativity of the modders, the altruism of the correspondent bankers, the community-mindedness of the Tribunal judges, the humility and inventiveness of PARO caretakers, the canny friendliness of the small-scale Let's Play creators who reach out to the isolated and lonely among us. It is possible to glimpse in these instances of computer-mediated participation some wisdom and purity, some moments of perfection, and maybe even some kind of immortality that may be naturally part of good acts, whose reverberations we can never completely know.

11
The Governance of Social Life: A Story of "Work" and Institutional Implosion

11.1 Introduction

The widespread adoption of heteromated labor not only affects individuals differentially in the ways we discussed in the last chapter, it also has long-term social, economic, and political implications. Here we focus on the implosion of some of the institutional support structures—social welfare, health insurance, pensions, and retirement—that constituted the social safety net in the United States and beyond for many decades. These institutions of support were the outcomes of struggles and developments in the first part of the twentieth century that gave rise to the so-called welfare state. The reversals in capitalist economies launched in the 1980s provided the beginnings of the unraveling of the welfare state (see chapter 3). What we explore here is continuous with that process and, in a deep sense, its logical outcome. We call these phenomena "implosions" because what seems to be happening is the undermining of some of the key institutions of the welfare state from *within*, emptying them of the structures and resources that sustained them for many decades.

Heteromation is central to these processes, first through the cultivation of a certain type of citizenship and social membership, and then through the promotion of the deregulated, ungoverned environment desired by dominant capitalist ideologies of our time. Heteromation, in other words, provides an effective set of "technologies of government" (Foucault 1986). What neoliberal agendas had sought to accomplish by fiat, heteromated technologies are delivering on a silver platter, but with attendant predicaments that are typically discounted in those agendas.

Perhaps more importantly, the expansion of heteromated labor into many arenas of the economy is reshaping and redefining the concept of "work." Work, in the sense of waged labor and employment, constitutes a key institution of capitalist societies both in economic *and* social terms. Work constitutes a major component of both *living* and *making a living* for modern human beings. Who we are has a lot to do with what we do as "workers" (and also activities in the negative space of "non-work"). Heteromation is transforming notions of work and worker, along with what it means to be employed, unemployed, underemployed, or self-employed, changing people's self-image as well as social image along the way. We would like to examine these transformations as well.

11.2 Heteromation and Citizenship: The Personalization of Risk

The transformation of the ideal subject, described in chapter 3, is an ongoing process closely coupled with large-scale transformations of the socioeconomic system. It was through these coupled processes that the passive, prudent, and loyal individual of the late nineteenth century became the entrepreneurial, risk-taking, competitive person of the late twentieth century, who in turn has retained and deepened these characteristics to become the actively engaged and highly stimulated individual of contemporary capitalism.

These transformations enabled, and were in turn enabled by, larger schemes of social welfare and support. Whereas industrial workers of the first era were expected to attain insurance for themselves and their family members through trade associations, friendly societies, or, later, private insurance, by the mid-twentieth century in most Western countries protection was attained through citizenship, via national insurance plans.[1] The exception was the United States, where, starting roughly in the 1950s, private health insurance became more or less widely available. It was provided by employers or purchased by individuals (but there was no national plan, and many people went without insurance[2]). This time period was the dawn of the development of more extensive private offerings through "personalized" plans that encompassed a whole range of benefits, from health, accident, disability, and life insurance to pension, retirement, and unemployment protection. More recently, along with these plans, which continue in the United States today, each year a host of new gadgets is introduced in dazzling technology showcases to "assist" people with home security, health tracking, childcare and eldercare monitoring, and other services. These parallel transformations at the individual and social levels have given rise to what O'Malley (1992) called the "new prudentialism."

This contemporary prudentialism, however, is different from its nineteenth-century predecessor in a number of ways. Rather than being geared toward mutual support, protection against risk and uncertainty, lifelong stability, and social responsibility, this prudentialism feeds, in good neoliberal style, on personal autonomy, anxiety, lifestyle maximization, and individual responsibility. In this fashion, as Miller and Rose (2008, pp. 100–101) point out, "a virtually endless spiral of amplification of risk ... continually open[s] the construction of new problems and the marketing of new solutions." As a result, they argue,

> [I]ndividuals, educated through the mechanisms of marketing and the pedagogies of consumption and lifestyle, are to gain access to previously "social" benefits such as educational advantage, health status, and contentment in old age through purchase in a competitive market.

And here is where heteromation comes in. Heteromated technologies of the type we have examined in this book ceaselessly engage individuals in activities that instill and cultivate the kinds of attitudes, skills, and behaviors called for in the environment Miller and Rose describe, i.e., autonomy, competitiveness, stimulation, and total engagement. The tensions generated by such anxiety-driven prudence feed, in turn, back on the society, depriving people even further of much needed support structures.

Think of insurance pools such as auto or health insurance, for instance, and the effect that "personalized risk" might have on them. In the last hundred years or so, we have learned to accept some degree of uncertainty and error in our social arrangements. We have accepted the inability to separate those at the giving end (the cautious driver or the healthy individual) from those at the receiving end (the reckless driver or the person who is sick)—hence appreciating the social necessity of some level of mutual support. With individual pricing of insurance, however, this might turn out to be a historical anomaly for the United States, dissipating our willingness to share some risk and social burden with others (Ohm and Peppet 2016).

Obamacare faces this challenge, largely from neoconservative pundits and politicians who try to convince "the young and the healthy" that they should not be paying for the coverage of the elderly and the sick. The business model of Health IQ, which we discussed in chapter 10, is that "people who take responsibility for their health are overpaying for a number of financial products because they are subsidizing those who are less Health Conscious."[3] Such arguments, despite their shortsightedness, are a reflection of changes in notions of citizenship and (lack of) solidarity in the current environment. As Croll (2013) put it, these are "… our generation's civil rights issue[s], and we don't know it."

11.3 Heteromation and Deregulation: The On-Demand Economy

Another set of institutions threatened by the expansion of heteromated labor includes those that have provided some measure of security and stability to the large segment of the population that lives on waged labor: pension and retirement funds, labor and trade unions, childcare benefits and facilities. This development is also continuous with the policies and practices of the last few decades in which organizations have sought to replace work processes involving context-specific knowledge of employees by the formalized, routinized, and "automated" work of outsourced labor, which can be shipped out to low-wage markets around the globe (Schiller 2014, Silberman 2015). As we argued in earlier chapters, however, in reality these so-called automated systems could actually work only with the involvement of new kinds of human labor that are less visible and hence less rewarded. The story does not end here. For this kind of labor to be delivered, many other things have to shift: management becomes "algorithmic" (Irani 2013), employment turns into contractual relations with much lower income to the worker, and other established benefits are eliminated in the process.

To understand these shifts, we can start with technology itself. McAfee and Brynjolfsson (2011), in their discussion of "the second machine-age," provide two examples of such technology, a close examination of which is enlightening.

11.3.1 Baxter: The Legless Humanoid Robot

The first example is the robotic system called Baxter, developed by Rethink Robotics, the stated mission of which, as the company name implies, is "to pursue and build *un*traditional industrial automation"—e.g., robots that can pick up and place jelly jars on an assembly line. This is a task that was hitherto performed by human beings because jars are delivered to the line in cardboard boxes, and "traditional" robots do not have the dexterity to take the jars out of the boxes and put them on the assembly line. Given the repetitious character of the task, it is indeed a laudable goal to eliminate it from human labor—a goal that apparently inspired Rodney Brooks, the former director of MIT's AI lab and the current owner of Rethink Robotics. Brynjolfsson and McAfee (2011, p. 13) describe the approach as follows:

> Brooks envisions creating robots that won't need to be programmed by high-paid engineers; instead, the machines can be taught to do a task (or retaught to do a new one) by shop floor workers, each of whom need less than an hour of training to learn how to instruct their new mechanical colleagues.

Adding that Baxter is an inexpensive ($20,000) humanoid robot with LED eyes and claw-like grips for hands, the authors observe, "It doesn't have legs, though; Rethink [Robotics] sidestepped the enormous challenges of automatic locomotion by putting Baxter on wheels and having it *rely on people to get from place to place*. The company's analyses suggest that it can still do lots of useful work without the ability to move under his [sic] own power … To train Baxter, you grab it by the wrist and guide the arm through the motions you want it to carry out" (ibid., emphasis added).

This description does not tell us which one of the two tasks the shop floor workers might prefer to perform: taking the jars out of the box or pushing robots around and training them literally through handholding. Nor does the story tell us what type of "analysis" led the company to the conclusion that the robots' lack of locomotion was not an issue, despite the fact that they "aren't as fast or fluid as a well-trained human worker at full speed." The concluding remarks by Brynjolfsson and McAfee, however, provide some clues:

> Baxter has a few obvious advantages over human workers. It can work all day every day without needing sleep, lunch, or coffee breaks. It also won't demand healthcare from its employer or add to the payroll tax burden. And it can do two completely unrelated things at once; its two arms are capable of operating independently.

"Demanding healthcare" and "needing sleep" become accusations of a sort. Although Brynjolfsson and McAfee are not heartless by any means, they slip into normalizing somewhat uncaring rhetoric as they enumerate Baxter's "advantages."

11.3.2   Kiva: Amazon's Army of Mini-Robots and Minimum Wage Labor

The other robotic technology Brynjolfsson and McAfee (2016, p. 13) discuss is one designed by a company called Kiva, which was acquired by Amazon in 2012 for $700 million:

> … Like metal ottomans or squashed R2-D2s, [these robots] scuttle around buildings at about knee-height, staying out of the way of humans and one another. They're low to the ground so they can scoot underneath shelving units, lift them up and bring them to human workers. After these workers grab the products they need, the robot whisks the shelf away and another shelf-bearing robot takes its place. Software tracks where all the products, shelves, robots, and people are in the warehouse, and orchestrates the continuous dance of the Kiva automatons.

Again, the narrative does not tell us what the workers do when they grab the products—for instance, how they lift the potentially heavy items from these low-lying ottomans. More importantly, it doesn't tell us that Amazon treats these workers as contractors managed by a temporary staffing agency paying them minimum wage with no benefits, or that the contract workers can be dismissed without notice if they are sick, if they miss a day of work, or if they do not meet the performance criteria set and monitored by computer systems (O'Connor 2013a,b).

Amazon, of course, is not alone in this kind of practice. Rather, it represents a bigger trend toward the "sharing economy"—or more aptly, the "on-demand economy" (Kessler 2015)—embodied in companies such as Uber, Lyft, Handy, TaskRabbit. The appeal of this trend is so strong that older companies such as FedEx have tried to emulate it in their employment practices (Rooney 2014). Stone (2006) captures the trend in three key terms: flexibilization (flexible employment of the sort we described in chapter 3), globalization (relocation or expansion of work to low-wage parts of the globe), and privatization (the dismantling of the social safety net).

Rather than a simple neglect, the absence of these topics in Brynjolfsson and McAfee's account derives from a kind of thinking that treats human beings as pieces in computing systems, putting "products, shelves, robots, and people" all in the same group, and investing hope in software to "orchestrate" their interactions. Like many other commentators of computer technology, the authors take their cues from technologists, entrepreneurs, and venture capitalists such as Rodney Brooks and Remi El-Ouazzane, the former vice president of Texas Instruments, who holds "a firm belief that the robotics market is on the cusp of exploding" (Brynjolfsson and McAfee 2011, p. 13).

To appreciate the full range of the impacts of these technologies, however, we need to also look to the trenches, where low-paid shop floor employees, contract workers, and Amazon Mechanical Turkers face the realities of the new unregulated environment on a day-to-day basis. We would then understand why the "recovery" of the economy in the aftermath of the 2008 Great Recession has created more low-paying jobs in place of the high-paying jobs that were lost in the recession. In light of this situation, the prognosis offered by some observers of computer technology about what the future holds for human beings, then, becomes quite damning:

> The main lesson of thirty-five years of AI research is that the hard problems are easy and the easy problems are hard ... As the new generation of intelligent devices appears, it will be the stock analysts and petrochemical engineers and parole board members who are in danger of being replaced by machines. The gardeners, receptionists, and cooks are secure in their jobs for decades to come. (Pinker 2007, pp. 90–91)

The not-so-long-term effect of these trends, beyond declining wages and the shrinking of the middle class, is the depletion of social funds (pension, retirement benefits, and so on) that have provided the wage-earning population with some degree of security in the past. A recent demonstration of this was the passing in 2014 of the Multi-Employer Pension Recovery Act (MPRA), which allows for significant cuts to the pensions of those who are already retired and on a fixed income. Attempts to reverse this act under the Keep Our Pension Promises Act (KOPPA) have proven futile thus far.

11.4 Heteromation and Work: Job Readiness, Social Worthiness

These shifts bear on some of the most fundamental aspects of modern life, particularly "work." What does it mean to "work" in an on-demand economy? Who is a worker and who is not? Who is employed, who is unemployed, and who is self-employed? What types of responsibilities, rewards, and benefits accrue to each? What types of laws and legal protections apply? How do people perceive themselves, and how do others perceive them, in each category? These are questions that impact the majority of people in contemporary societies, not only in the economic sense but also in the sociocultural and psychological sense. For work is not just a way of making a living, it is also a way of living, closely intertwined with all dimensions of life from family and friendship to freedom and wellbeing.

And because it is so central to modern life, there are many perspectives on the meaning of "work" and its relation to these other dimensions. These perspectives understand differently the current shifts in economy, technology, and their relationship. In particular, the perspectives would have distinctive evaluations of heteromated labor and its relationship to work. A proponent of the on-demand economy, for instance, who sees the current situation as people "monetizing their own downtime," speculates as follows:

> [W]e are defining a new category of work that isn't full-time employment but is not running your own business either ... We may end up with a future in which a fraction of the work force would do a portfolio of things to generate an income—you could be an Uber driver, an Instacart shopper, an Airbnb host and a TaskRabbit. (Manjoo 2015)

This sentiment is also shared by the economist Alan B. Krueger, a former chairman of President Obama's Council of Economic Advisors, who highlights the "flexibility" and higher income of working in the on-demand market as a source of its appeal—a claim that is strongly disputed by Robert Reich, the former US labor secretary, who, through his own studies, characterizes this work as "unpredictable, doesn't pay very well and is terribly insecure" (ibid.). "Can you imagine," Reich speculates:

> if this turns into a Mechanical Turk economy, where everyone is doing piecework at all odd hours, and no one knows when the next job will come, and how much it will pay? What kind of private lives can we possibly have, what kind of relationships, what kind of families? (ibid.)

Both of these perspectives for and against the on-demand economy, however, start from the assumption that work is inevitable—an assumption that originates in the capitalist idea, going back to thinkers such as Adam Smith (see chapter 2), that work is a necessary activity for the production of wealth. A very different perspective emerges if one questions the validity of this assumption, as Kathi Weeks does in her book, *The Problem with Work*. Weeks is perplexed by the widespread acceptance of the present reality that one must not only work to live, but live to work, and asks, "Why do we work so long and so hard?" To answer this question, she starts with the observation that:

> Work is crucial not only to those whose lives are centered around it, but also, in a society that expects people to work for wages, to those who are expelled or excluded from work and marginalized in relation to it. (2011, p. 2)

This brings Weeks to the notion of work as a social convention with great moral pressure, a disciplinary apparatus that dominates our lives, and a means of subordinating the worker to the employer and the capitalist.[4]

Weeks's observation sheds some light on current debates on the on-demand economy and the future of work, particularly on the relation between heteromation and modernist notions of work. Understanding work as a moral, conventional, and disciplinary mechanism reveals a number of couplings that are usually ignored or underplayed in discussions of work: the coupling between work and employment, between work and social (and self-) image, between work and equality, and between work and freedom. Different perspectives on the future of work can be actually understood as playing up some of these couplings and playing down others. Using the rhetoric of "flexibility," the proponents of the on-demand economy highlight the separation of work and employment to infer that contractual workers are relieved from the burden of fixed hours, repetitive tasks, and direct supervision. Liberal-minded critics, on the other hand, stick to the tight relation between work and employment to emphasize the "right to employment," while their right-wing neoconservative rivals push for the "right to work." A radical critique such as Weeks's, however, understands these "rights" as various incarnations of the same modernist/capitalist viewpoint, bringing to light the close connection between work, subordination, and inequality.

These various plays on the relationship between work and other aspects of modern life—autonomy, freedom, and equality, on the light side, and domination, subordination, and exploitation, on the dark side—reveal some of the capacities of heteromated systems in the long run. Within the capitalist system, heteromation has the potential to decouple many of the mentioned pairs, except for one. It can decouple work from employment, or from direct domination, but it *cannot*, in our estimate, undermine the extraction of value by a small minority from the labor of the majority. Here is the secret to its mystery.

The average person, all the while, has to not only struggle to decode this mystery, but must also adjust to the reality of a world that still evaluates persons largely on the basis of their relation to work without giving them any guarantees of work.

12

Utopias: A Story of Revolution and Reform

12.1 Introduction

The light-and-dark character of heteromation provoked some soul searching on our part, and we wonder about emancipatory changes that could reconfigure heteromation from marginalized labor to equitable employment, relaxing the moral imperative of certain forms of paid labor as a necessary part being a good person (discussed in chapter 11). Now we contemplate a particular vision of a future of work, and of life more generally, by sketching a utopian design. As we have seen, much heteromated labor is freely chosen, enjoyable, interesting, challenging, and rewarding in a surprising number of ways. At the same time, heteromation, with its low returns to workers, is a driver of economic inequality.

Some reject utopias as nothing more than fantastical scenarios. But we agree with sociologist Erik Olin Wright that "real utopias" are essential imaginaries for discovering ways to define and enable real change.[1] Drawing on eclectic sources, we propose a real utopia, followed by comparison with proposals that are more reformist in nature. We move beyond heteromation, considering the economy as a whole, in fact, pushing all the way to the planetary limits that fundamentally constrain economic activity (see Georgescu-Roegen 1971; Hornborg 2014). Underlying the economic inequality Piketty so thoroughly documents is the despoliation of the very Earth itself, and the resources upon which all human economic activity, and indeed all human life, depend. Capitalist projects are set in motion so that the rich can discover how much richer they can get, an absurdist endgame of accumulation by dispossession, exploitation, heteromation, and other forms of accumulation.[2]

Considering the magnitude of the problems this accumulation has caused, we turn to the work of steady-state economists who theorize the economy as part of the global ecology. These economists insist that we cannot proceed intelligently until we acknowledge the finiteness of Earth's resources. Herman Daly (1991) and Douglas Booth (1998), in particular, examine how we might bring economic activity into balance with the Earth's carrying capacity. Culture and society can grow and develop in unlimited fashion, they say, but resource use and waste accumulation cannot. The laws of thermodynamics prevent that. Booth (1998, p. 156) observes:

> As the economic system expands, it places increasing demands on the global ecosystem for energy, materials, and ecosystem services. The global ecosystem, however, has a fixed capacity to provide such services (Daly 1991). As a consequence of economic expansion, the global ecosystem suffers from excessive exploitation, and, as a result, its capacity to provide inputs to the global economy is diminished.

Contemporary Marxist thought also includes ecological approaches (e.g., Burkett 2003), and while there is disagreement about the role of natural resources and ecological processes in the dynamics of capitalism, Hornborg (2014) points out that squabbles notwithstanding, the ecological approaches stand with the subdiscipline of ecological economics and proponents of a steady-state economy in demanding an accounting of natural resources in theories of economy. It is neoclassical economics, Hornborg wryly observes, that "does not seriously consider material constraints on economic processes," (2014, p. 91)[3]. The neoclassical paradigm is, needless to say, the dominant paradigm, the 800-pound gorilla compared to which the other approaches are small if scrappy contenders. In this book, we have emphasized labor as the foundation of economic value, a touchstone necessary to explain the development of contemporary digital technology. But we recognize, with scholars such as Burkett and Hornborg, that analysis of the Earth's resources is indispensable to any broad theory of economic activity. The realities Booth delineates frame the discussion.

This chapter constitutes something of a small manifesto. It might turn out that the ideas will be of little interest to readers whose primary goal in reading this book is to think about heteromation as a computer-mediated labor relation. We wrote the chapter, however, because critiques such ours often evoke the response: What is to be done? We do not, of course, have simple answers to this question. But we consider how the kinds of labor that are currently heteromated could be embedded in more equitable labor relations. We push questions of labor out to a larger context, exploring proposals for a guaranteed basic income, new forms of subsistence, and a true sharing economy. We believe that the disciplinary pressure to work that Weeks questions might be altered by implementation of some of these proposals.

12.2 André Gorz and *Paths to Paradise*

We begin back in the 1980s, when Austrian philosopher André Gorz wrote a small but insightful book sketching a utopia that would address technology, economy, and inequality. In *Paths to Paradise*, Gorz (1985) envisioned a more relaxed economy and society. This society would reverse the accumulation of wealth and power into the hands of an ever-smaller group of elites, and give everyone some room to breathe and live. Gorz welcomed automation, suggesting that we could reconfigure the economy to favor the masses through the use of technologies to relieve workers of certain forms of burdensome labor, and reduce time spent at paid labor. Release from long hours of work, and from debilitating or stultifying labor, would be traded off for less consumerist ways of living. Gorz reckoned that with low-consumption lifestyles, and in the company of robots, most of us could work about five months a year. Weeks asked why we work "so long and so hard." Perhaps the answer is that we do not necessarily have to.

A life with less labor in the 40-hour-a-week model of disciplinary control would be a big change from the way employed Americans typically work.[4] If we ceased producing the extraordinary volume of cheap and obsolescent junk that ends up in landfills, and reduced mobility and other energy sinks, we would not need to rouse ourselves to spend 40 stressful hours a week in the factories and call centers and warehouses that rob us of time with our children, the elderly, the infirm, and, indeed, with ourselves. A new cell phone every year is not part of a program of Gorzian simplicity, nor are extravagant levels of international travel, nor driving large vehicles to the grocery store. A basic guaranteed income *is* in the program, as well as a state that safeguards basic rights, including education and healthcare.

Gorz was adamant that the status games and aggressive competition woven into the texture of contemporary capitalism must stop if we are to build a saner society. He noted that status seeking stimulates consumption and the desire to accumulate wealth. With unchecked competition, the objective must become to drive everyone else out. Increasingly, contemporary capitalism does just that. Wealthy, powerful oligopolies (Suarez-Villa 2015) have gained even more control than at the time Gorz was writing. Competition does not occur on the level playing field mainstream economics imagines—rather, a small number of corporations exerts massive influence over media, the legal apparatus, and government (ibid.).

Countervailing forces to inject socially aware values into discussions about the society we hope to have are diminished when there is control by the few (Suarez-Villa 2015; see also Stiglitz 2014). The damage democracy suffers in these circumstances is appalling, as Morozov (2011), Suarez-Villa (2015), Brown (2015), and others have discussed.

Suarez-Villa (2015, p. 252) reports that oligopolies are accompanied by an astonishingly lopsided distribution of household net worth:

> The top one percent now has more private net household wealth, or net worth—the value of all assets possessed, after taxes—than the bottom 90 percent of the population, a situation that is almost unprecedented in the US. Only in periods preceding major crises, such as the 1920s, or in socially distressed nations—the kind that are usually thought to be ruled by oligarchies—can similar statistics be found.

Thomas Piketty has written at length about the funneling of wealth into the hands of a smaller and smaller number of very wealthy people (2014). He addresses inequality with proposals for redistributive mechanisms of global corporate taxation—which, as he notes with admirable good cheer, have been derided as utopian. While his proposals lack the sweep of Gorz's notions of refashioning society from top to bottom, Piketty's ideas constitute an important critical device for slicing into the complexities of what is going wrong in today's capitalism. The ideas are not mere procedures, but percipient critiques with ramifications beyond tax policy. Most important, they urge us to ask questions such as "Why don't we have these tax policies now?" They make visible a complex set of concerns we must begin to address.

Considering the realities these scholars (and others) have documented, the alternative to articulating paths to paradise and designing real utopias is to fold our cards now, yielding to powerful interests that would dictate practically everything. Such a course of inaction on our part seems reckless—in catastrophic ways, capital's markets have not worked. Huge numbers of people remain impoverished in the world today. We are caught in an unmistakable and terrifying downward spiral of environmental devastation. In the United States, home of the least regulated system of capitalism, forty million people live below the poverty line. Over two million are incarcerated. Mortality rates are increasing, a shocking reversal after decades of improvement.

A working market should not produce these outcomes. The objective of society—a path to paradise in broad outline—should be to shift the capitalist emphasis on growth in material throughput to emphasis on the growth and development of human persons and cultures. Such a society would not tolerate the outcomes we have mentioned. Petcou and Petrescu (2014, p. 264) clarify that the needed transition is not to be thought of as "sustaining" the current system, but as addressing "societal change and political and cultural reinvention … [including concerns of] inequality, power, and cultural difference."

How can such a transition be achieved? In the following dicussion, we propose some ideas, and then critique some others.

12.3 Revolution: Recovering Subsistence

Moving beyond concerns of today's economy, we consider historical trajectories of civilizational rise and fall from which we might grasp priorities for action. Archaeologist Joseph Tainter has studied these trajectories across time and space. Using the detailed, precise methodologies of archaeology to examine a range of historical societies in his sweeping book *The Collapse of Complex Societies* (1988), Tainter shows that all civilizations eventually collapse, declining over a period of decades or centuries. For rich countries, decline will result in less material abundance as we push the limits of the Earth's resources necessary for economic activity. Many poor countries, or at least large populations within them, may be said to have already entered decline, or even collapse.[5] Eventually, we will turn away from capitalism as it is practiced today because the

Earth cannot sustain its excesses (Daly 1991, Booth 1998, Vandermeer 2011, Robertson 2012, IPCC 2013).

But it is not necessary for our society to end in abject collapse. The societies that Tainter studied—the Maya, the Mesopotamians, the Minoans, the Inca, the Romans, the Egyptians, and others—did not possess the resources of science, history, and technology that we have amassed in the last 500 years. These resources have the potential to be usefully deployed to fashion a transition from the current, unsustainable system, to a new system based on designs for real utopias. There is no guarantee we can accomplish such a transition, but it seems likely that we have some time to plan, at least within typical Tainterian time frames.

But where to start? Probably we should begin with the means and the mode of subsistence production. Wright (2009) observes that capitalism initiated its domination by dispossessing workers of their means of production. Harvey (2003) draws attention to capitalism's strategy of accumulation by dispossession in which workers' once-autonomous activities are (forcibly) brought into circuits of capital—for example, economic upheavals have driven people all over the world from subsistence farming to cash cropping. As we enter the shaky political economy arising within a destabilizing natural and social environment, it seems prudent to disrupt such processes, finding ways to once again control of our own subsistence. Even if we reject the moral imperative to work that Weeks discusses, basic needs must still be met through some form of labor.

Most fundamentally, we need to eat, regardless of anything else that is happening in our lives. Surely, however, we are long past thinking about producing our own food? Current environmental trends indicate that food production is, in fact, something we should absolutely be thinking about. Air and water pollution, soil erosion, the destruction of fisheries, and precipitous drops in populations of wild species such as bees to pollinate food crops, are rapidly degrading the natural resources upon which all food production depends. These externalities, from which no part of the globe is spared, put food security in peril. Processes of resource degradation result from modes of agricultural production that depend on nonrenewable resources—fossil fuels, chemical fertilizers, herbicides, and pesticides. These substances are not only nonrenewable, they generate massive ecosystem effects such as polluted waterways, oceanic dead zones, and 29 percent of the world's greenhouse gas emissions (Vermuelen, Campbell, and Ingram 2012).

In California, to take but one instance, rice farming in the Sacramento Valley deploys aviation as a core technique: "Flying at 100 mph, planes plant the fields from the air" (Fox 2015). In addition to the fossil fuel for flying planes, rice production requires enormous amounts of water in a region of low precipitation. All of this, and yet most of the rice is exported, and few workers are employed (Fox 2015). The objective of such industrial agriculture is to generate wealth for the few. Scarce public resources (water allocations) are deployed to create this wealth, but the wealth is not distributed to the populace through employment.

Hornborg (2014, p. 92) observes that the increasing crop yields stoking optimism about feeding the world's billions "have primarily been based on imports of guano, phosphates, oil, and other resources from extractive sectors of the world economy." These resources, extracted from finite supplies, cannot sustain agriculture in the long run. Transferring the practices and machinery of this mode of production to poorer nations, instead of innovating ecologically sensitive techniques and employing local labor, is primarily a means of capital accumulation as corporations seek new markets for their products. By contrast, research on informated techniques of food production such as "computational agroecology" (Raghavan et al. 2016) and "printable gardens" (Takeuchi 2016) promises to provide some of the necessary innovation for a return (albeit a technologically inflected one) to at least some subsistence food production.

Basic subsistence also requires the manufacture of everyday tools, household items, shelter, transport, and so on. Recent computing research has addressed means by which we might begin to recover autonomy of production. A philosophy of "DIY" informs work on "making," hacking, craft, undesign, repair, and reuse, signaling a desire to counter massification and restore some autonomy.[6] "Maker" culture (Ratto 2011, Bradley 2014) promotes local production using 3D printing to fabricate clothing, shoes, implements, machines, craft materials, and much more.

These dual efforts in computer-mediated subsistence agriculture and DIY manufacture are attempting to discover ways to return fundamental processes of production to the masses, that is, to *distribute the means of production*. Capitalism distributes primarily consumption. The extent to which a viable system of distributed production is possible remains to be seen. Co-optation is always a threat, including powerful interests that would heteromate the labor involved in these efforts. We believe, though, that the developments we mention have the potential to recover individual and small group production of daily essentials, reversing, at least to some degree, accumulation by dispossession, exploitation, and the intensification of heteromation.

It seems unlikely, however, that *everything* will be non-market—wealth must be generated and circulated to sustain civilization. If it is not, we will end up in small, poor, isolated, regional societies. We might, for example, produce a piece of train track on our 3D printer, but to lay track from point A to point B, a larger, cooperative, regulatory unit (supported by taxation) must exist, or we cannot connect the links to build an infrastructure.[7] Advances in human rights, such as greater acceptance of LGBTQ individuals, result from social movements scaled at national or international levels (see Nardi 2013). For such reasons, it does not seem desirable to restrict society to small units or to go overboard on localizing production or governance. Tainter (1990) observes that collapsing civilizations result in exactly such small, regional units, which eventually collapse or are conquered by more powerful societies with superior organization.

In realist Gorzian fashion, then, we envision that labor beyond subsistence will go to government and corporations.[8] Together, the state and corporations, properly managed, could produce basic services such as railroads, schools, healthcare, goods that cannot be 3D-printed, and so on. Would these institutions be managed any better than they are today? We believe that if they are not, the environmental collapse inherent in current economic practice will lead to chaos. As Tainter documented, most societies eventually collapse. We optimistically assume that with advances in science and philosophies of human rights, we have some chance of finding ways to transition to a system more like the steady-state economy Herman Daly has envisioned.

Interrupting processes of capitalist accumulation would relieve the pressures of too much work for too little reward (Gorz 1985, Nardi 2015). In concert with a resumption of subsistence activity, a basic guaranteed income would contribute to a non-negotiable safety net. This safety net requires taxation and well-regulated corporate activity. The basic guaranteed income, also known as "social income" or "universal basic income," is being tried in Finland, the Netherlands, Canada, and elsewhere. Historically, it has many precedents; for example, in the United States between 1862 and 1986, fully 10 percent of the land mass was given away by the government for homesteads.

The social income has demonstrated its appeal to at least some members of both the right and left sides of the political spectrum. On the right, the social income upholds libertarian values, rejecting the nanny state which attempts to anticipate and supply specific needs. The social income advances a bracing "40-acres-and-a-mule" mentality: individuals are given basic resources, and they figure things out for themselves from there. The state need not concern itself if a citizen decides to play video games all day and live on a small income. It's too expensive and intrusive to find out what the citizen is doing (see also Brynjolfsson and McAfee 2016).[9]

Over on the left, a basic income policy declares that no wealthy nation should allow any citizen to live in dire poverty, and everyone should be guaranteed a minimal living. Conditions such as homelessness, for example, are unconscionable in this view.[10] The social income, then, gives people the means to act responsibly for their own well-being in neoliberal fashion, yet guarantees that no one falls over the economic edge. Milton Friedman liked the social income because it dismantles expensive means-testing bureaucracy. Martin Luther King liked it because it favors a more equitable society. We are thus hopeful that such a solution could take root, given the political bedfellows who have, over the years, championed various forms of this type of wealth distribution.

The extent to which capitalism would have to change its game as we tread a path to paradise is unknown. Contemporary, winner-take-call capitalism is not playing the long game. Its lack of care for the planet, disregard for the billion among us who go hungry, and tolerance for increasingly endemic global economic precarity make it as shortsighted as a quarterly report.

However, certain favorable portents indicate that capitalism *could* be moved toward less avaricious versions of itself. Alternative styles of corporate management, including employee-owned corporations and benefit corporations, have already established a firm legal basis; they do not need to be invented or legislated. Such corporations could reduce the trend toward oligopoly, while leveraging capitalism's special nimbleness and bottom-up creative ferment. Schemes like global taxation that keep the lid on concentrations of wealth (such as those Piketty suggests) seem essential if some form of capitalism is to be viable. Or a new system must emerge. Capitalism organizes society—nothing more and nothing less—and it can be changed or even abandoned.

In suggesting a return to control of subsistence, we want to be clear that we are not proposing a romantic attempt to recover lost connections to nature or a simpler, preindustrial society. That would be preposterous. No one wants to go back to a world without obstetrics or recorded music or washing machines. We must discuss a real utopia that combines elements of previous social systems with new elements such as the internet and robotics. In particular, the legacy of exploitation of living bodies manifest in agrarian societies is incompatible with modern ethics. That exploitation is no longer needed—it is now possible to insert technology into more equitable and environmentally sensible programs of economic activity. The healthful physical activities of subsistence farming might be encouraged, and we might develop trimmer, stronger bodies as a result, but we need not return to the arduous labor that once wore peasants out at an early age and would prevent participation of the elderly and disabled. An interesting technical challenge, for example, would be to develop technologies such as robots for subsistence agriculture that engage a variety of participants at differing levels of physical robustness.

Now we must discuss some of the practicalities of a real utopia, examining new materialities of production and distribution and the role of formerly heteromated labor in sustaining the parts of civilization worth keeping. The solutions we propose either exist now or are present in current research agendas, although some of them are rather unusual.

12.4 Materialities of Production and Distribution

12.4.1 Agroecology

Most of us know little about growing our own food. Gardening and smallholder farming have hardly died out, however, and efforts such as permaculture and perennial polyculture (Ferguson and Lovell 2014) are increasingly relevant. The scientific study of agroecology is concerned with creating "self-sustaining, low-input, diversified, and energy-efficient agricultural systems" through techniques that promote biodiversity (Altieri 1999, p. 30; see also Vandermeer 2011). Instead of fossil fuel–based industrial farming techniques, agroecological systems utilize ecological principles and expertise about local conditions. Unused capacity for sustainably growing food in private and public lands can be leveraged to shift a significant portion of food production to techniques such as perennial polyculture that support cultivation of fruit and nut trees, grapevines, berry bushes, and perennial food plants like artichokes, asparagus, leeks, rhubarb, and herbs. We can envision a wide diversity of environments, from urban/suburban smallholders to larger farms full of productive, sustainable food-producing ecosystems.[11]

To fulfill this vision, new technology is necessary. Agroecological practice is complex and takes many years, sometimes decades, of hands-on experience to master. It relies on precise knowledge of local conditions, including "biogeochemical conditions, climatology, plant, animal, and insect species, topography, soil ecology and chemistry, agroforestry, water management, inter- and intra-specific plant competition, terraforming, sunlight requirements, and plant propagation" (Raghavan et al. 2016). Climate change creates the need for continual new knowledge. Industrial agriculture, by contrast, depends on brute force inputs of fertilizers, herbicides, and pesticides. It is a more standardized, portable methodology, but one with increasingly expensive externalities.

Computing systems that extract, systematize, and distribute knowledge of bioregional conditions and practices of cultivation have the potential to make agroecology scalable (Raghavan et al. 2016). Diverse formal and informal systems have the necessary knowledge, but this knowledge is scattered and inaccessible. Communities of local growers, for example, collaborate and exchange information online, and their knowledge and support could be scaled up to encourage people to learn about ecologically sensible food production. Copious data from government databases and the scientific literature could be curated and made accessible (Raghavan et al. 2016). Takeuchi (2016) addresses the same problem of scaling subsistence agriculture, offering technologies based on 3D printing. Both computational agroecology and printable gardens have the masses in mind, seeking to develop practical ways for anyone to grow food, even those with little agricultural knowledge.

12.4.2 Making, Sharing, and Repairing

The intensive consumption that began in the era of massification has run its course, at least from an environmental perspective. We must provision ourselves with basic material goods, but the finiteness of the earth's resources indicates that we cannot keep treating its services (topsoil, clean water, minerals, fossil fuels, animals such as bees and bats) as though they are inexhaustible. Recent critical thinking in the field of human–computer interaction including sustainable HCI (Blevis 2007, Brynjarsdóttir et al. 2012, Knowles et al. 2013, Dillahunt 2014, Joshi and Cerratto-Pargman 2015), undesign (Baumer and Silberman 2011, Pierce 2012), and repair (Maestri and Wakkary 2011, Jackson 2014), has upended taken-for-granted assumptions about the inevitability and desirability of contemporary consumerist lifestyles. In "simple living" groups (Håkansson and Sengers 2013), for example, people choose thrift and simplicity, creating lifestyles and practices that could inform plans for broader societal changes. Ecovillages re-skill residents in methods of food production and shelter fabrication (Nathan 2008, Cerratto-Pargman, Pargman, and Nardi 2016). A recrudescence of the 1970s social movements of voluntary simplicity and appropriate technology are visible in such efforts. Steampunk devotees practice reuse, repair, and recycling in their communities. Notably, they engage in constant "critical reflection on the role of technologies" in society (Tanenbaum, Desjardins, and Tanenbaum 2013, p. 109).

Products designed to be repaired interrupt cycles of planned obsolescence, a phenomenon first observed in the 1950s by Vance Packard (1960). Repair decreases profits but promotes environmentally sound practices and the self-reliance that underpins workable systems of subsistence production. These approaches indicate that "motives of consumption," as Marx called them, are social productions—they are not "natural" or given. We should see them as such, and consider how they can contribute to environmental sustainability and the provision of basic necessities for all.

Isn't the "sharing economy" gearing up to help us gain more equitable economic structures, lowering barriers to participation and distributing wealth? In the case of companies such as Uber, the sharing economy seems a way to push risk to workers, as Robert Reich (2015) has explained. The millions of dollars in venture capital that underwrite such enterprises belie the intent to "share."

But a different, genuine sharing economy is emerging in practices of computer-mediated freecycling, barter (Knowles et al. 2013), and time-banking (Bellotti et al. 2014). These projects implement peer-to-peer solutions for redistributing wealth outside markets (see Bradley 2014). Freecycling can be as simple as a neighborhood listserv where people announce their castoffs, and neighbors drop by to pick them up. In many cases of peer production and distribution, the technologies exist, and it is a matter of people developing practices for using them.[12]

We believe urges toward usage of such tools is growing. Even where technical solutions are not yet in place, people are expressing strong interest in innovative means of sharing within communities. For example, in Japan, there is a local currency community in the city of Ueda in the Nagano Prefecture that organizes economic and social activity around the invented local currency, the *maayu*:

> [The residents] use a *maayu* bankbook to keep a record of exchanging services and goods. Exchange using the local currency includes, for example, taking care of pets while the owner is away, guiding people in mountain vegetable hunting, fixing the handles of kitchen knives, giving rides, trimming trees in yards, and selling home-grown vegetables, homemade wine, and unneeded wedding gifts [!] … When this group created their own local currency, they also started a market called "*maayu ichi*" (*maayu* market) in order to have a place for members to meet face-to-face once a month. There, participants bring things they want to sell, including homegrown vegetables and homemade meals, and they eat together occasionally. … Through this and other projects, members became better acquainted with each other and the community became solidified. (Ueno, Sawyer, and Moro 2016)

Gui and Nardi (2015) documented similar sharing activities in a town in Devon, United Kingdom, where people organized workshops to share knowledge about gardening, fermenting vegetables, repairing bicycles, beekeeping, and the like, as well as undertaking larger projects such as building and repairing houses for low income residents. As in Ueda, the social activity and neighborliness that blossomed from the economic activity were as important to residents as the practical services (ibid.).

The postwar years of massification and related developments discussed in chapter 3 interrupted attitudes informed by an ethos of DIY/sharing/self-sufficiency—attitudes that had been commonplace in earlier generations. As the economy began to rebuild after the war, the recovery eventually entered a kind of overkill that promoted a fierce consumerism, as capitalism, and its attendant culture, lost all sense of balance. Powerful interests refused to take seriously the dislocations of frenzied economic activity which were eroding environmental and social sanity. Those dislocations continue to plague us today. Yet it is a hopeful sign that DIY sensibilities were not stamped out entirely. We find them still, experiencing a revival in cultures as varied as Japan, the United Kingdom, and the United States.

12.5 Where We Are Now

Perhaps Marx would not be surprised at the problems that concern us today. Caffentzis (2013, p. 237) recalls Marx's observation that social systems based on money entail heavy costs "from depressions, famines, and slavery to police, prisons, and execution chambers to banks, stock markets, and all sorts of expensive 'financial services'." With contemporary phenomena such as volatile stock markets, escalating police violence, and untrustworthy banks and lenders, Marx's observation is more timely than ever. Reducing dependence on these institutions seems prudent, sensible, even necessary. The costs of current markets with their relentless monetization of nearly everything, and the avarice that results from the cocooning of the extremely wealthy from the effects of their activities, are rarely taken into account when examining the so-called "rationality" of markets.

At present, the new (or revived) materialities of production and distribution we have discussed are scattered and piecemeal, occurring in little islands of enthusiasm. Their value lies in the knowledge they generate, which can be used now and in the future. This chapter, as a mini-manifesto, has emphasized proposals for actions that do not require organized social movements with agendas easily defeated through corporate-funded legal action.[13] We have instead drawn attention to activities that can be carried out by individuals and small groups that nonetheless constitute substantial, practical responses.[14] Even if the tax schemes Piketty proposes (and other such reforms) never materialize, there is still much to be done in our own backyards, basements, and community spaces.

We take inspiration from rapid, mass turnarounds such as the Victory Gardens cultivated during World Wars I and II[15] which produced almost half of all consumed fruits and vegetables during the war years. Nearly instantly, these efforts transformed food production, as Americans were inspired to grow food for themselves so that commercial production could go to soldiers and the starving populations of Europe (Hayden-Smith 2014). Rapid change *is* possible. Producing our own food is possible. And so is making our own everyday items. When we can print customized objects using open-source software without corporate mediation, perhaps we will do so, just as people only one or two generations ago possessed skills in sewing, knitting, carpentry, woodworking, leatherworking, machine repair, and blacksmithing. (Production is, of course, easier with a 3D printer.) We note that practices of winemaking, tool repair, and gathering wild foods are still present in Japan.

It is perhaps ironic that in this age of massive networks, individual and small-group efforts loom large as key methodologies of change. The neoliberal economy continues to actively perturb traditional forms of mass resistance. In the United States, for example, recent legal decisions have disempowered unions, once a formidable source of resistance. TTIP (Transatlantic Trade and Investment Partnership) and TPP (Trans-Pacific Partnership) will shift even more power to corporations. Finding ways to circulate new ideas and practices outside these structures, which stack the deck against all but the wealthiest, is a challenge before us.

It seems, though, that we might have at least one trick up our collective sleeves, and that is to turn neoliberal ideology on its head by taking our "freedom" and using it to learn subsistence techniques, to educate ourselves about peer-to-peer methods of production and distribution outside markets, and to adopt open-source software alternatives that empower us to use the technology for our own ends.[16]

These activities, of course, require computing machinery and digital networks. It would be possible for powerful interests to withhold, or to price us out of, for example, the rare earth minerals required for computing, or to otherwise manipulate the system. Capitalists are well organized in robust organizations such as the IMF and the World Bank. Select heads of state and global corporate managers jet to meetings at venues such as the World Economic Forum where class interests are given an annual airing. There is indeed reason to be concerned that existing powers would attempt to derail the sorts of activities we have been proposing.

However, ideas for countermeasures also exist, and we must pay attention to them, because if we do not maintain a level of optimism and practical engagement, paralysis is likely. With respect to the all-important project of sustaining widespread access to the internet, Raghavan and Hasan (2012, p. 1) discuss what they call an internet "quine"[17]—a design for an independent, self-reproducing internet:

> The Internet stands atop an ... industrial system required for its continued growth, operation, and maintenance. While its scale could not have been achieved without this reliance, its dependencies—ranging from sophisticated manufacturing facilities to limited raw materials—make it vulnerable to disruptions. To achieve independence requires an Internet quine—a set of devices, protocols, manufacturing facilities, software tools, and other related components that is self-bootstrapping and capable of being used by engineers to reproduce itself and all the needed components of the Internet. In this paper, we study how such an Internet quine could be built. We also attempt to identify a minimal set of such tools and facilities, and how small and inexpensive they can be made.

The authors conclude that internet technology relies on widely available resources, with a few important exceptions such as gallium (for semiconductor fabrication), which occurs in sizeable deposits only in Australia, Brazil, Kazakhstan, and Venezuela. Despite potential shortfalls of such minerals, Raghavan and Hasan (2012) point out that we have manufactured so much computing equipment already, that for the foreseeable future it will probably be possible to build what we need with salvage.

Raghavan and Hasan observe that a quine could interoperate in a flexible manner with the current internet (although performance might be less than what we are used to). They note that hobbyists have been manufacturing radios for decades, and we could do the same with the internet. More broadly, research on "computing within limits" (Pargman and Raghavan 2015) is beginning to examine how we can maintain technical infrastructure with some independence. For example, Patterson (2015) reported that intermittent energy availability in Haiti has led end users to take more control of infrastructure, thereby changing labor relations and social life.

12.6 Transitioning from Heteromation

Following Gorz, the utopia we have been sketching requires a state and a corporate sector (the two eventuating, we hope, in something like a steady-state economy). Within state and corporate sectors, some of the types of tasks that are currently heteromated could provide employment. A new labor relation would be necessary, one that does not marginalize workers, but would provide a fair return for labor and a share in governance. This labor would no longer (by definition) be heteromated—heteromation would wither away, shifting the kinds of work it now organizes to a more equitable basis. It is the labor relation that needs changing, not the labor itself. As we saw with modders, citizen scientists, graphic designers, writers of political commentary, video producers, and so on, the work may be engaging—the problem lies in inadequate compensation, protections, and participant control.

Computer-mediated labor could materialize as what Alvin Toffler began to envision in 1980 as the "electronic cottage." Microwork in systems such as Mechanical Turk, were it to return a fair wage and offer worker protections and worker inputs to decision making, has a place in a future economy. Working at home on a flexible schedule and performing modular tasks could benefit those engaged in childcare or eldercare, the disabled, the chronically ill, and others for whom full-time employment outside the home is undesirable or impossible. Such arrangements require a social income so that workers are not driven to self-exploit and can enter employment relations in a relaxed, productive manner. We believe the economy will continue to be organized around computing for many years (or centuries) to come, and that formerly heteromated labor will be needed, providing useful employment. Kathi Weeks and Bob Black questioned the need to work under regimes of capitalist discipline to achieve moral standing—we suggest that compensated, computer-mediated labor might begin to alter the idea that a 40-hour-a-week job is the basis of propriety and respectability.

A positive development in this regard is the ILO's report on work-on-demand which argues that "casualization and informalisation of work and the spread of non-standard forms of employment" require labor protection: "[F]undamental labour rights to all workers irrespective of employment status" are essential (de Stefano 2016, p. 111). De Stefano points out that vested interests would like to categorize on-demand work as something other than actual work (as fun, or as "extra" income that supplements a real job), but that that is hardly the case. The ILO takes a firm stand on treating on-demand workers as employees entitled to benefits such as a minimum wage (de Stefano 2016), a hopeful sign pointing the way to a sensible society—perhaps even a society on a path to paradise.

12.7 Reform

Since the 1980s, when the brief postwar golden age of responsible capitalism ended, moderate, reformist proposals for mitigating the dislocations of capitalism have been put forward, and, to some extent, mainstreamed. These proposals are genuinely critical of contemporary capitalism and dissect the problems insightfully. But, at the same time, they seem hesitant to deeply imagine change.

One key reformist influence is the work of sociologist Amitai Etzioni, who perceived the hard edge that capitalism was developing in the 1980s, and in response, wrote a book, *The Moral Dimension: Toward a New Economics* (1988). Here Etzioni proposed "communitarianism" as an alternative basis for society—a way to alter the ugly, competitive mood that seemed to be driving out common decency. Communitarianism would shift employment from corporate enterprises to volunteer work, and to work in non-profit and nongovernmental organizations. These venues, Etzioni argued, could ground our lives in "moral," noncommodified social relations.

Though well intentioned, this proposal puts the onus on the masses to become "better" people, living their lives within a "moral" universe of nonprofit and volunteer work. Burdens of change are placed on the shoulders of the people, who are given little real power, with the ruling classes more or less let off the hook. Caffentzis (2014, p. 265) argues that communitarianism defeats the logic of its own critique—it is more likely to brace up capitalism than to take away the hard edge:

> [N]ongovernmental organizations inspired by [communitarianism] have rushed into the various catastrophes caused by neoliberal structural adjustment policies around the planet (from Detroit to Somalia) to save "humanity." But, in this process, they have also helped save "the market" and, by the same token, the very policies that allowed for the development of such catastrophes.

Propping up the regime, though far from communitarianism's hope, is the likely outcome of urging people toward humanitarian efforts responsive to, and ultimately complicit with, capital's blunders. The discourse of morality nudges us toward choices such as assisting with short term disaster relief that appeal to our better natures, but that do not transform the material basis of everyday, practical life. Until we have control over our basic needs we will always be chasing capitalism's tail (and ultimately we will have to reckon with capitalism's demise from which the rich will have plans to shield themselves).

A second reformist influence is Jeremy Rifkin, whose ideas on work have developed over many years and are presented in his books *The End of Work* (1995) and *The Zero Marginal Cost Society* (2014). He proposes a future in which capitalism recedes, giving way to a "sharing economy" that will bring "free" things to all. Rifkin (2014, p. 21) says:

> [M]illions of prosumers are freely collaborating in social Commons, creating new IT and software, new forms of entertainment, new learning tools, new media outlets, new green energies, new 3D-printed manufactured products, new peer-to-peer health-research initiatives, and new nonprofit social entrepreneurial business ventures, using open-source legal agreements freed up from intellectual property restraints.

Nothing could seem more revolutionary than "free everything," delivered from productive collaborations taking place in a plethora of different activities within the zero marginal cost society. Yet Rifkin leaves capitalists in charge of automated agriculture, manufacturing,[18] and service industries such as insurance and banking. How free will those goods and services be? Rifkin shifts attention from the intricacies of a real economy to his notion of a sharing economy, arguing that capitalism will not hold a candle to the plenitude and innovations of the Commons, which he illustrates with a stream of disconnected examples.

Though Rifkin believes in free stuff and the power of the masses to create a Commons-based economy, his cases are often constructed on contradictions. For example, he reports that research on 3D printing is financed by "the U.S. Department of Defense, the National Science Foundation, and NASA" (Rifkin 2014, p. 96)—hardly bastions of the Commons. These powerful institutions are not likely to gracefully fall back when their role in funding the creation of technologies such as 3D printing is complete, nor will the resultant technologies fail to bear the imprint of the ideologies and interests of these organizations.

Rifkin predicts, probably with justification, that in the future the majority of workers will have been replaced by machines. He suggests shifting labor to a "third sector" between the public and private sectors—that is, shifting to non-profit and volunteer work! It is difficult to see how there could be *that* much volunteer and non-profit work to do, yet both Etzioni and Rifkin argue that this is the way forward. Gorz's proposal that we work five months a year at societally necessary labor (teaching, construction, healthcare, plumbing, and so on) seems to make a good deal more sense.

Viewed critically, the notions of "sharing" and "community" that frame Etzioni's and Rifkin's schemes appear to conceal a bigger picture that is rather terrifying. In moral economics and the zero marginal cost society, capitalists control the core sectors of the economy, while workers are shunted to jobs creating new forms of entertainment, helping out at disasters, and so on. Capitalists, managing their machines with a small number of workers, make decisions about food, manufacturing, and essential services. Ceding basic economic functions to capitalism and marginalizing workers to volunteer and non-profit sectors would seem to encourage capitalism to continue down the road of environmental irresponsibility and social inequality it is currently traveling. Rifkin (2014) remarks that climate change threatens food security, but he seems not to connect this observation to the automated agriculture he would have capitalists manage.

"Prosumers" in the Commons can only do what they do if their subsistence is in place, either through their own production, or through the portion of a wage that is necessary labor-time (as discussed in chapter 3). Assuming we do not return to a completely subsistence-based economy, necessary labor-time must come from a job or a welfare transfer based on someone else's labor. Can the non-profit, volunteer realm provide this labor? Caffentzis (2014, p. 73) points out that this is, in fact, impossible: "The capitalism resulting from Rifkin's 'new social contract' [i.e., the nonprofit third sector] is impossible, for it is by definition a capitalism without profits, interest, and rents."

Rifkin's vision entails another mystification involving the millennial generation's putative capacity for "empathic engagement" (Rifkin 2014, pp. 281, 284). Millennials, says Rifkin, naturally choose cooperative, Commons-compliant actions organized around the Internet of Things. Rifkin's zero marginal cost society emerges as a strange brew—empathic engagement; technologies that somehow appear, exogenous to capital's interests; free goods and services; a Commons without visible means of support; and capitalism without profits.[19]

A third reformist influence is the work of economists Piketty (2014) and Brynjolfsson and McAfee (2011) whose careful research reveals important current and future trends. Piketty drives home critical points about inequality, demonstrating that wealth is accumulating in fewer and fewer hands. The argument is solid and essential, but reformist in proposing taxation as the key remedy (as important as this remedy is). This approach does not acknowledge the larger global complications in which we are mired. Piketty has obviously been asked what the thinks about the world's big problems, and he devotes a few pages to climate change in his book, admitting that climate change is "the world's principal long-term worry" (2014, p. 567). But he says only that "no one knows how these challenges will be met" (p. 567). Of course, Piketty makes no claims that he has answers. Yet we cannot but hope for broader perspectives and further wisdom from one of our most astute observers.

Brynjolfsson and McAfee's analysis (2011) shows that automation is reducing the number of conventional jobs, and that there is no obvious sector to absorb the surplus labor, as there has been in the past. Their findings are particularly useful in countering the "new jobs have always come along" argument (see also Gordon 2012). However, despite the rigor of their analysis and their advocacy of a social income (Brynjolfsson and McAfee 2016), their main proposal is quite limited, i.e., better education to address massive job losses. But what jobs are we to be educated for? MOOCs and similar solutions cannot do much if the number of conventional jobs requiring education continues to shrink. As technology gets smarter and the pool of labor becomes increasingly global, it will be more and more difficult for workers to win the "race," as Brynjolfsson and McAfee call it, even with superior education.

Graphic designers are a case in point. A cohort of elite, sophisticated workers, they compete in design competitions they cannot win, even with impressive education and experience. There's no dismissing the graphic designers as impractical, artistic types when we consider that software development, once thought an impregnable fortress of employment, is being heteromated through the development of techniques of crowdsourcing (LaToza et al. 2015). Indeed, it seems that fewer and fewer jobs promise what we now consider normal employment. Hence, we affirm the sanity of Gorz's proposal to drastically reduce the amount of time a person spends performing the labor required for the remaining necessary jobs that support society.

A fourth reformist influence is the writing of commentators such as Naomi Klein (2014). These commentators are deeply concerned about the ill effects of contemporary capitalism, but their proposals for change tend to look back, not forward, and are psychological in nature. Klein, for example, suggests that fundamental change requires refiguring narratives and worldviews. Specifically, the masses should adopt attitudes reflecting ecological attachment to place (as in traditional societies) and they should participate in local economic activity. While Klein's exposition of the problems of capitalism is forceful, her notions of change are atavistic rejections of modernity and industrialism. She finds hope in indigenous people's struggles against industrial capitalism. But these are disconnected, doomed movements of the dispossessed. She says, a bit sentimentally, "When what is being fought for is an identity, a culture, a beloved place that people are determined to pass on to their grandchildren, and that their ancestors may have paid for with great sacrifice, there is nothing companies can offer as a bargaining chip" (2014, p. 342). But cultures all over the world have been decimated by capitalism precisely because they have no bargaining chips. The Dark Mountain Collective in the United Kingdom espouses similar views, proposing that we need new "stories" about how to live. Their writing and artwork celebrate nature and places far from the grit of cities and the bleakness of poverty-stricken rural areas.

We are in sympathy with these perspectives, which spring from concerns very similar to our own, but their nostalgia ends up unintentionally obscuring the role of the the ruling classes in creating the problems.[20] We think it more likely that we can devise ways to use digital technology in an end run around at least some aspects of capitalism by producing our own subsistence and looking to methods of peer distribution of goods and labor. We can try to shade capitalism down into more civilized manifestations, such as corporations owned by employees or focused on societal benefits. And we can reform taxation as Piketty suggests.

Is this a "new story"? It doesn't seem like it's new; people "did for themselves" to a great extent prior to World War II, and taxes on the rich then (and into the 1950s and 1960s) were much higher. We think what is needed is to scale up pragmatism and common sense. The planetary destabilization contemporary capitalism increasingly produces feels like the antithesis of common sense.

12.8 Back to (Part of) the Future

Drawing widely from Gorz, Marx, and recent research in computing, we have proposed a somewhat radical vision of transformative change. We do not conceive of "overthrowing" anything, but rather of detaching certain key functions from capitalism, moving them under our own control to interrupt capitalism's frenzies. We have proposed a real utopia that modifies the socioeconomic system in three ways: (1) it recovers control of subsistence through digitally mediated food production, local fabrication, and computer-mediated means of distribution; (2) it implements a social income; and (3) it moderates work arrangements to reduce work time and offer the option of labor in the electronic cottage at fair wages. These changes would avoid the deficiencies of most reformist proposals, which accept (perhaps unconsciously) that the sunk costs of current capitalist structures cannot be abandoned and are to be left in place. Our vision aims to begin to reverse the consequences of accumulation by dispossession, exploitation, and intensified heteromation, a move we regard as essential for serious social and environmental renovation.

This vision of a real utopia, though we call it "revolutionary," is not really so far-fetched. After all, when capitalism began, methods of accumulation were far less entrenched, and their impacts less widespread and intense. Braudel's (1979) descriptions of early capitalism in Europe during the fifteenth to eighteenth centuries reveal a great number of varied arrangements, at least some of which both sustained moderate profit and secured worker autonomy and fair wages. A balance between waged work and production for family and community emerged, even if it only occurred by accident, and even if it only occurred under certain historical conditions. Workers who tended their gardens and animals, and earned modest wages through paid work can perhaps be aspirational for us as we evolve new ways of using computing, as well as more socially responsible forms of socioeconomic organization. The objective is to mediate processes of production and labor with a greater degree of fairness and with a view to future generations.

Epilogue: The Story of Machines and Us

Computer technologies, like many other technologies of the past, are not simple—technically, but also in regard to their relations to us. Digital technologies have a strongly ambivalent character—they are empowering, liberating, and transparent, yet at the same time intrusive, constraining, and opaque. We enjoy a sense of empowerment and emancipatory self-expression afforded by digital technologies, but anxiety and confusion perplex us in the face of the rapid changes they have entailed, not all of which we like. There is a light side and a dark side, and we find it hard to tease them apart—probably, most of the time.

For these reasons, and perhaps many others, our seemingly simple question about the changing relations between labor and computing did not yield simple answers. The question invited us to engage in certain moves toward theory, drawing on the ideas of giants such as Marx, Foucault, and Weber, along with many other thinkers and writers, to bring the ideas together with our own understanding of computer technology. Yet these engagements seemed to always end in new questions. It was a bit like a game of whack-a-mole—with every question to be addressed by the theories and our experiences, another popped up. We found ourselves taking one "detour" after another. Here we want to take you through these detours, and highlight insights from each, before coming back to the original question.

Detour 1: Personal Doubts and Social Demands

One of our very early questions (even before we had homed in on heteromation as a focus) concerned how we contend with doubt and confusion in the form of questions and anxieties punctuating everyday life. Am I spending too much time playing this video game? How can I get my spouse to stop using his smartphone obsessively? Shouldn't I go take a walk? Am I unreasonable to feel this annoyed at that person and her loud cell phone conversation? How genuine are my experiences in the digital environment, and what "good" is attained through them? How can I justify (dis)engagement with this environment to myself and to (ir)relevant others? How often should I check my email, and how quickly should I respond to text messages? Should I sign onto Facebook, LinkedIn, Twitter, and the many other sites thrust at me every day? How do they affect my professional life and career? Do they enhance my opportunities for connection, employment, and advancement, or do they undermine my privacy, security, stability, and time for other pursuits? Who should my friends be? And how often should I check their posts? How should I feel about my followers? What should I share with them, and what should I hide? Should I take "technology vacations," as some pundits advise, or even "digital detox"? In brief, we wonder if digital technology is improving our well-being as individuals and societies, or changing things in ways that are neither in our interests nor within our control. Is digital technology taking away the personal, communal, and human aspects of our lives, or enhancing them?

Despite such questions, most of us, most of the time, find ourselves attached to and engaged with digital technologies. They seem inevitable, indispensable, and irresistible; they invariably evoke fascination, attraction, and strong desires for their offerings. Certainly the technologies enable much of value: information gathering, sharing opinions, political participation, connecting to friends, family, and colleagues, crunching numbers, checking the weather, analyzing texts, listening to music, playing games, taking care of ourselves, and a multitude of leisure pursuits. At the same, we observe others in incessant (obsessive?) engagement with the technologies, inviting us directly or indirectly to do the same.

We notice an increasing number of government, business, and social services (such as healthcare) that can only be accessed digitally. It is strictly within the computerized environment that we can operate as "normal" members of our communities and societies. Even critical commentators who urge caution about technology deliver dispatches from their digital devices, composing texts for transmission, via a few clicks, to the plentiful crop of readers they can reach on the internet. Pervasive systems such as computing appear inevitable, and we accept them as fate. Despite a certain measure of doubt, we are increasingly wrapped up in the technologies, accepting them with sincere appreciation for their utility and convenience, but also succumbing to their coercive power, and the peer pressure and social demands they seem to bring forth.

Technologies appear to come and go according to an inner logic beyond human control. This kind of thinking, that technology is inevitable (in whatever form it currently happens to be), is often referred to as "technological determinism," and has been the subject of scrutiny and criticism for several decades. Critics, coming from traditions such as the social constructivist perspective, contend that this view ignores the human element in the development and adoption of technology. Ultimately, they argue, it is human beings who are in charge, making choices about technology—a proposition that makes intuitive sense once made. Despite such criticism, however, the deterministic argument has had great staying power, not dissimilar to mythical dragons. In fact, every time its critics think that they have cut off its head, the Dragon grows it back with ever more resilience.

Detour 2: Magical Machines and Helpless Humans

To understand the determinism implicit in the core of many current accounts of the relationship between computing and economy, we need to interrogate a more basic relationship—namely, that between humans and machines. This relation is often understood in terms of a naturalistic and essentialist perspective that attributes inherent properties and capacities to humans and technologies: *what each is good at*. The intuitive appeal of the language of humans-are-good-at-certain-things and machines-are-good-at-others makes it the more difficult to challenge the thinking behind it.

We saw one example of this in chapter 1 in the views of influential figures such as Herbert Simon and Richard Langlois. The false predictions that Simon made about the future of computers—e.g., "machines will be capable, within twenty years, of doing any work a man can do" (Simon 1960)—have revealed some of the fallacies in his thinking. Brynjolfsson and McAfee (2011, p. 62) take note of Simon's claim, and argue that, "the set of tasks machines can do is not fixed. It is constantly evolving. ..." (2011, p. 62). This understanding, however, does not stop them from presenting the following thesis:

> [T]here's never been a better time to be a worker with special skills or the right education, because these people can use technology to create and capture value. However, there's never been a worse time to be a worker with only "ordinary" skills and abilities to offer, because computers, robots, and other digital technologies are acquiring these skills and abilities at an extraordinary rate. (p. 3)

This view leads the authors to attribute feats to technology that give it much more power than it actually has: "Today's information technologies favor more-skilled over less-skilled workers, increase the returns to capital owners over labor, and increase the advantages that superstars have over everybody else" (ibid., p. 74). With technology being a key driver of this "skill-biased technical change," as Brynjolfsson and McAfee call it, it would seem that the only option left to human beings is to adapt to it (although even this option is suspect, as, according to the authors' own logic, humans are chasing the moving target of increasing machine competence, a competence that is "constantly evolving"). In this fashion, the socioeconomic developments of the last few decades, including the increasing inequality that concerns Brynjolfsson and McAfee, and many others, come to seem to be only natural outcomes of technological developments.

One implication of this logic is to put the blame on the doorstep of those who have been the losers in the playing out of these trends, giving the "losers" but one option: educate and equip yourself with the right set of technical skills, and you will have a chance to be saved. Humans are invited to play a game of catch-up with powerful machines. Or at least those humans with access to elite educational institutions are invited. In their continued invocation of the idea of "superstars," Brynjolfsson and McAfee do not seem to have any suggestions for the rest of us, except that we should try and educate ourselves about technology, presumably because that is what computer technologies "favor."[1] The naturalized view tends to attribute too much agency to machines, leaving humans at the mercy of technology and making invisible the power relations that underlie the drive to automation, heteromation, and the other "inevitable" economic moves made in the name of capitalist growth. To counter this tendency, and to understand the drivers of change, we need to unnaturalize both humans and machines.

Detour 3: Dynamics of Change, Statics of Statistics, and Traps of Psychologizing

If it is not technology *only* that drives change, then, what else influences the growing socioeconomic disparities of the last few decades? The answer to the question obviously depends on one's attitude toward these disparities—i.e., whether they are fair, natural, and sustainable or unjust, anomalous, and dangerous. We have neoliberal economists such as Larry Summers, the former U.S. Secretary of the Treasury, who celebrates inequality, considering it "the other side of successful entrepreneurship ... that is something we surely want to encourage."[2]

Many of us do not, in fact, want to celebrate or encourage inequality. There is little disagreement about the growing inequality of the last few decades. Economic statistics, which provide the basis for almost all interpretations, are quite telling in this respect. According to the latest Oxfam report, for instance, between 1988 and 2011, 46 percent of overall income growth around the globe went to the top 10 percent of the population, while the bottom 10 percent received only 0.6 percent (Oxfam 2016).

To see why this state of affairs is unfortunate, consider the view, shared by people on both sides of the political spectrum, that in the last few decades, our societies have transitioned from a normal distribution of income, wealth, and social mobility to a power law distribution (Oxfam 2016). The common explanation starts with the assumption that people appear on the social scene with varying degrees of "motivation," which then puts them on different tracks with respect to their level of participation in economic activities, ultimately determining their share of the economic pie.

The problem with this entrepreneurial, psychologized theory of human behavior, as we discussed in chapters 9 and 12, is that it puts the psychological cart in front of the socioeconomic horse. In so doing, it place the burden of change on individuals, sanctioning the manipulated "nudging" of their behaviors through delicate mechanisms of control. An alternative approach is to look for the drivers of change in inherent tensions of modernism and capitalism—a tack that students of Marx, Foucault, Weber, and others have pursued. Accounts based on this approach demonstrate patterns of cyclical development, where one tension gives rise to the next, one social settlement desettles another, and one opportunity undermines the previous one. Our formulation of this approach in terms of predicaments and possibilities is intended to capture this dynamic with an eye to the interactions between computing technology and the capitalist economy. This formulation led us to a "rewriting" of the history of computing (chapter 1), of capitalist change (chapter 3), and of the drivers and mechanisms of engagement with computing technology (chapters 4 and 10). To understand the growing inequalities of the last few decades, we must look beyond statistical trends to identify the underlying drivers of change that determine its dynamics.

Detour 4: Inside or Outside of Capitalism?

Let us now return to the problem of harmony in the capitalist system, as professed by neoliberal thinkers. Capitalism is often referred to as a "system" because, like any other social arrangement, it involves a large number of parts (people, institutions, technologies, and so on) interacting with each other in all manner of political, economic, and cultural relations. And systems have an "inside" and an "outside," as any systems theorist would postulate, which brings up the question of who/what is inside capitalism, and who/what is outside. Although we rarely ask this question, there is a common and implicit assumption that we are all *inside* the capitalist system because we are all *part* of it. This, of course, makes intuitive sense because we all, in fact, operate within the system, playing by its rules, benefiting from its resources and institutions, and contributing to it according to our ways and means. By this logic, anyone who lives in a capitalist society is an insider, regardless of their status, class, race, ethnicity, gender, education. There is a great deal of truth to this thinking, but there is also much more to the simple logic than meets the eye.

To see why, consider the case of Native Americans (or any other aboriginal people, for that matter) during the European takeover of the continent in the fifteenth and sixteenth centuries and afterward. The colonial system that drove this process—another "system" with its own parts and its own inside and outside—did not consider Native Americans as belonging to it. The natives were outside the system; there was "them" (the savages, people without civilization), and there was "us" (the civilized, the superior). Thus, European settlers in North America, by and large, oppressed the natives—hence the repugnant folk expression of the nineteenth century: "the only good Indian is a dead Indian" or, earlier, the Catholic Church's easy acceptance of enslavement of the indigenes as a legitimate social formation. European settlers and missionaries, on the other hand, were part of the colonial system regardless of how filthy, brutal, or murderous they were. That they were the "subjects" of the monarchs of England or Spain automatically gave them status as insiders. This division of peoples was a relatively clear-cut case of a system with an inside and outside.

The story of the capitalist system, however, is less straightforward. Take a nineteenth-century lower-class Londoner of the kind described by Dickens (say, Oliver Twist), or an early twentieth-century factory worker, such as the one portrayed by Charlie Chaplin in *Modern Times*. These individuals were inside the capitalist system, not only in the sense that they were the "subjects" of a sovereign power, but in the sense that the capitalist system needed them. And because it needed them, it provided them with at least two things: a means of subsistence and a justification for why they should play along with capitalist rules. And since these provisions were not always adequate, there was also a need for mechanisms of control. In this fashion, control and consent are historically and closely tied to each other, as Marcuse (1964) formulated (see chapter 3). Early on—that is, roughly up until the mid-twentieth century—control took a rather direct form, demanding obedience from workers on the shop floor. The good worker of industrial capitalism was, therefore, not a dead one, or a slave, but an obedient person.

The difference between the "good Indian" and the "good worker" is important in that it can help us understand the basis of class exploitation in early capitalism. With this difference in mind, Erik Olin Wright (2005) formulated a theory of class exploitation on the basis of three criteria:

1. *Inverse interdependent welfare principle*: The material welfare of exploiters causally depends on the material deprivations of the exploited.
2. *Exclusion*: The exploited are excluded from access to certain productive resources.
3. *Appropriation*: Exclusion enables exploiters to appropriate the labor effort of the exploited.

These three conditions, in other words, should be in place in order for a relationship to be considered exploitative. That is why, according to Wright, the industrial worker, but not the Indian, can be considered exploited; Indians were oppressed but not necessarily exploited (apart from the slaves of the Spanish).

How about the rest of us—all the Googlers, Facebookers, Twitterers, YouTubers, Instagrammers, Mechanical Turkers, gamers, citizen scientists, self-service customers of banks, insurance companies, and other corporations, and the rest of the army of billions that we selectively accounted for in part II of the book? Are we exploited?

On the one hand, we deeply agree with those who identify a strongly negative, exploitative, and coercive thread in the way computing technology supports and feeds current capitalism. On the other, we could not fail to notice, and partially share, the sense of fascination, excitement, and optimism that surrounds this technology. It was in dealing with this intellectual predicament that we came up with the notion of heteromation and the whole conceptual apparatus that accompanies it here.

The concept of heteromation, for us, strikes a meaningful and pragmatic balance between these views in a number of ways. First, it does justice to the labor, ingenuity, and creativity of the growing number of human beings who contribute value in the current economy, gaining no or minimal reward, recognition, or compensation. Second, it recognizes the power of computing technology and its transformative potential for socioeconomic, cultural, and political change, while avoiding the common deterministic fallacy of putting humans at the mercy of machines. Third, it provides a fair understanding of the sociocultural mechanisms that drive participation and engagement, but that also obscure the delicate techniques of control and even coercion in the current environment. In doing so, it corrects the psychologizing tendency of attributing undue credit and blame to individual motivations and shortcomings, and puts the responsibility instead where it indeed belongs: the unsustainable structure of the current capitalist system. Fourth, heteromation captures and integrates the economic and social aspects of the predicaments of contemporary life at both the individual and collective levels. It can, as such, unveil some of the mystery that surrounds recent technological developments and socioeconomic displacements that accompany them, and that leave many in doubt and darkness about their current and future lives. These mechanisms largely work through a logic of inclusion (*not* exclusion, as in Wright's explanation of exploitation), drawing people into the fold of computing while at the same time keeping them out of the circle of capitalist elites or denying them a meaningful role in the governance of their lives, communities, and societies.

Detour 5: To Automate or to Heteromate?

This tension between digital inclusion and social exclusion creates a predicament for capitalism. At its base, the predicament is not new; it is something that Marx noticed a long time ago. As we discussed in chapter 3, Caffentzis (2013, p. 72) summarizes Marx's point:

> Hence, the capitalist class faces a permanent contradiction it must finesse: (a) the desire to eliminate recalcitrant, demanding workers from production, (b) the desire to exploit the largest mass of workers possible.

The predicament, in other words, derives from the fact that, on the one hand, capitalism needs human labor as the sole source of value and, on the other, it aspires to minimize or even eliminate the cost of labor. How could it, then, deal with this tension? It turns out that the tension cannot be easily resolved, as any serious economist would readily acknowledge. Consider Hal Varian, Google's chief economist, who, like many other theorists of current capitalism, recognizes the deep economic changes brought about by computer technology—in particular, the reduction of what economists call the "marginal cost of reproduction" to almost zero. What this means is that it costs almost nothing to reproduce and distribute information because of the properties of the digital medium (Ekbia 2009; Kallinikos, Aaltonen, and Marton. 2013). The production of information, on the other hand, is a very different story, as Shapiro and Varian (1998, p. 21) cogently describe: "Information is costly to produce but cheap to reproduce." This very statement captures an opportunity that economists have smartly identified: minimize production costs, and you have a good business model for a product. (See also Caffentzis's [2003] discussion of this point).

An example of such a product is the instantaneous online translation service Google Translator, which basically generates translations by mixing and matching the fragments of all human-generated translations that the system has in its vast repository of texts. The repository is the gold mine that Google digs into in order to provide its services. It is with gold mines such as this that Brynjolfsson and McAfee (2011, p.27) wonder, "What would happen to the digital world if information were no longer costly to produce? What would happen if it were free right from the start?" To this question, they candidly provide the following answer:

> The old business saying is that "time is money," but what's amazing about the modern Internet is how many people are *willing* to devote their time to producing online content without seeking any money in return. ... The billions of hours that people spend uploading, tagging, and commenting on photos on social media sites like Facebook unquestionably creates value for their friends, family, and even *strangers*. Yet at the same time these hours are uncompensated, so presumably the people doing this "work" find it *more intrinsically rewarding than the next best use of their time*. To get a sense of the scale of this effort, consider that last year, users collectively spent about 200 million hours each day just on Facebook, much of it creating content for other users to consume. That's ten times as many person-hours as were needed to build the entire Panama Canal. (ibid; emphasis added)

This description is hardly in need of commentary, except for a few "little" assumptions:

1. People *willingly* contribute their time and effort.
2. They find it, "intrinsically," more rewarding than the next best thing.
3. Their effort benefits even "strangers."

While it is difficult to dispute the first assumption on its own terms, we can understand what "willingness" really means in light of the second assumption. Imagine an unemployed, underemployed, or even an employed college graduate with job for which they are over-qualified, of which there are plenty nowadays, and consider what is the "next best" thing that they can do with their time. Or, consider a person, even a manager, with a relatively well-paying job, who comes home in the evening or on the weekend, tired and exhausted after working in a tense environment that demands more than the paid 40 hours without overtime compensation, including the expectation that the worker will stay online at home, lest she miss an email or a call from her manager. What is the next best thing that this individual can do with her time?[3] Such scenarios, all too common, do not seem to penetrate sanitized discourses positing rational economic actors calmly deciding among intrinsic rewards.

And, then, there are "strangers," whom Brynjolfsson and McAfee do not name, making us wonder who they have in mind: the neighbors around the block who have lost their jobs to automation; the small downtown shop owner on the verge of bankruptcy because Walmart has opened a Superstore in the area, the local coffee shop taken over by Starbucks, the elderly person who must struggle through long phone menus trying to fill a prescription; or perhaps the Congolese fisherwomen, graphic designers in Botswana, activists in San Salvador, and cattle herders in Serengeti who are so precisely invoked as beneficiaries of the new digital age by Eric Schmidt and Jared Cohen (2014; see Assange 2014, p. 54). We suggest that the "strangers" who benefit from people's free labor must decisively include the major corporations and other winners of new capitalism, some of whom we have discussed in this book.

Detour 6: Of Horses and Humans—and Machines

Our argument in the latter chapters of the book has been that the current state of affairs is not sustainable—socially, politically, or environmentally. This assessment is shared by many prominent academics and intellectuals, including economists (Piketty 2014, Stiglitz 2014), social scientists and environmentalists (Altieri 1999, Dyer-Witheford 1999, Vandermeer 2011, Bradley 2014, Hornborg 2014, Klein 2014, Reich 2015), religious leaders and humanists such as Pope Francis and members of the Dark Mountain Collective, and many others. These figures look at the current situation from very different perspectives and prescribe different solutions, to be sure, but they all share concerns about the risks that current trends pose to our civilization and to humanity as a whole. If this many credible people see the problems, then, how is it that the trend continues to expand with alarming recklessness?

The answer to this question is multifaceted, and we do not claim to even have the beginnings of an answer. We do, however, know that part of the answer is in how people, particularly elites with clout, act within the current situation to foster their own interests. Political, intellectual, and academic elites shape a great deal of the thinking and discourse about social issues. Although in this book we kept using the word "capitalism" as if it represents a personified unity, we are aware that in reality there is no such reified entity as "capitalism." What happens, rather, is that a set of processes has been set in motion in modern societies which have given rise to, among other things, what we call the capitalist system. Central to these processes is a "triumphant reorganization of capitalism that is deploying the new technological innovations to solidify an unprecedented level of global domination" (Dyer-Witheford 1999, p. 236). The inner dynamic of these processes is largely above and beyond the will of any single individual or group of individuals. The dynamic is earthly and material, however, and it provides a space for human intervention.

We have already seen one example of how intellectual elites might have influenced our thinking, and hence the course of events regarding computing and the economy, in the person of Herbert Simon—a towering figure who influenced twentieth-century thought in a broad range of areas, from economics, operations management, and organizational science to artificial intelligence, psychology, and cognitive science. We have also seen that, despite his many false predictions, Simon doggedly stood by his views. Now, to consider one of Simon's peers, let us quote another Nobel Laureate—Wassily Leontief—who made the following statement in 1983:

> The role of humans as the most important factor of production is bound to diminish in the same way that the role of horses in agricultural production was first diminished and then eliminated by the introduction of tractors.

Brynjolfsson and McAfee, who quote Leontief with approval, build on his proposition to argue for "technological unemployment"—i.e., driving human beings out of their jobs by technologies of automation.[4] We not only disagree with the premise of Leontief's argument about the diminishing role of humans in production, we also find the whole analogy between horses and humans misguided. We make this judgment not from an anthropomorphic and chauvinistic human perspective, but from a purely socioeconomic one. We consider statements such as the above as self-fulfilling prophecies of neoliberal thinkers such as Gary Becker (1962), who sought to create a new kind of subjectivity that reacts to situations in predictable ways. In their attempt to turn humans into predictable automata, these ideologues prophesied an image of humans as first akin to horses, and then to machines. What they failed to notice is the key difference between humans, other animals, and machines: it is not in what each is "good at," but in the degree to which their actions and behaviors can be regulated, monitored, and controlled. The same qualities that make humans malleable, shapeable, and, often times fallible, also allow them to reflect, resist, and revolt—that's what makes humans special, without equivalent in the animal kingdom or in the realm of machines.

Detour 7: The Slippery Slope of a Critical Perspective

We are almost full circle back to our original question about the complex and convoluted relationship of humans with machines. We cannot, however, leave the discussion here without some measure of self-reflection on our part. Our perspective in this book, which we have also pursued in our earlier works (e.g., Nardi and O'Day 2000, Ekbia 2008), can be broadly described as the "critical study of computing." We consider ourselves "critical friends" of technology. This line of research has a relatively long tradition in informatics and computer and information science, going back to the work of people such as Rob Kling (1996). The work has had varying degrees of impact on the real practice of computing, some of which might be more obvious and acknowledged than others. At the same time, the long tradition has also allowed and encouraged some degree of self-reflection and self-criticism within the community. A good example is the posthumous critique of Kling's legacy by some of his close collaborators (King, Iacano, and Grudin 2004). The authors admire Kling for, among other things, his "strong inclination to view as dubious any statement that was not grounded in empirical evidence or theoretical analysis, particularly those that encouraged people to take actions that would ultimately benefit those making the statement" (King, Iacano, and Grudin 2004, p. 3). Among such statements, the authors mention those made by the AI pioneer Marvin Minsky in an interview with *Life* magazine in November 1970:

> In from three to eight years we will have a machine with the general intelligence of an average human being. I mean a machine that will be able to read Shakespeare, grease a car, play office politics, tell a joke, have a fight. At that point the machine will begin to educate itself with fantastic speed. In a few months, it will be at genius level and a few months after that is powers will be incalculable. (Darrach 1970, p. 60; in King Iacano, and Grudin 2004, p. 2)

This and similar statements by Minsky and other AI practitioners interviewed by *Life* led the poor reporter to conclude: "Computers could free billions of people to spend most of their time doing pretty much as they damn please" (1970, p. 65; in King Iacano, and Grudin 2004). It was conclusions such as these, and the "expert" statements that provided fodder for them, that drove Kling, rightly and righteously, toward a critical perspective. And it was Kling's sharp and skeptical analysis that drew the admiration of others for his work. However, as King, Iacano, and Grudin point out, "a strong critical perspective can be non-reflective and too quick to dismiss other points of view," partly because of its "highly protectionist stance toward the common person and regaled against class advantages that technology might have for the rich and powerful" (King, Iacano, and Grudin 2004, p. 4). One outcome of such a stance, according to the authors, was Kling's failure to appreciate "the dynamics of exponential change." This is a kind of change that is, for instance, embedded in Moore's Law, and which the human mind is apparently incapable of easily grasping.[5] King, Iacano, and Grudin describe this as an outcome of "the slippery slope of a critical perspective."

The warning about the slippery slope cannot but give us pause about our intellectual vulnerability—a danger that we take very seriously. Although there is no potion or panacea that would protect us against this, we have tried hard throughout this book to avoid the slippery slope by maintaining a balance between the positive, productive, and promising aspects of the current computerized economy and its negative, dark, and damaging aspects. To that end, we have consulted a broad range of literatures coming from different perspectives, trying to do justice to their viewpoints. We have further drawn on our own empirical studies, and our own daily experiences, and those of those around us, to attain a concrete and grounded understanding of current circumstances.

Our labors have aimed not only to expose, to the best of our ability, the complex system of humans, machines, services, political entities, and economic interests that come together in multiple layers of mediation and remediation throughout our computerized environment, but to venture toward the equally complex issue of making evaluative judgments about technology. The computer-mediated activities we enjoy make a huge difference in our lives, in matters that are often and at once pragmatic, delicate, and deeply human in character. But this has come at the cost of inequality, insecurity, anxiety, and numerous other concerns that we have examined in this book. Yet seeds of change may lie within the very technologies we have discussed, although human reason, aided by our sense of fairness and mutual dependence, must prevail if we are to successfully cultivate them.

Notes

Chapter 1

1. See http://appleinsider.com/articles/10/12/06/apple_co_founder_offered_first_computer_design_to_hp_5_times.html.
2. See https://www.youtube.com/watch?v=-nz8WD6v-Jg.
3. See bricklin.com.
4. Schiller (2014, p. 39) notes that Eric Schmidt "affirmed before the World Economic Forum at Davos that jobs that had earlier seemed beyond the reach of automation now were endangered."

Chapter 2

1. Those who preceded "cliometrics"—the approach to economic history, originating in the 1960s, that applies the techniques of mathematical economics to the history of the field.

Chapter 3

1. In 2014, Southwest Airlines rebranded with the inadvertently Marxist slogan, "Without a heart, it's just a machine." The slogan was designed to acknowledge Southwest employees, and to reinforce the contributions of their strong customer support (which is, in our experience, superior to that of other American air carriers, exemplifying the contributions of living labor). Southwest understands that the airplane, i.e., the "machine" cannot sustain its business.
2. We differ from Caffentzis regarding why machines do not provide value. Caffentzis points to the "the ability of labor-power to refuse to be realized as labor whether within or without a contractual bond" as the separator. He invokes Melville's Bartleby, the famous scrivener who made the knowing choice to "prefer not to" when it came to his work tasks. Heteromation, though, is about the invisibility of crucial forms of labor, about people not recognizing the value of their labor and thus being unable to refuse. We agree that in many circumstances,

people *can* refuse labor, and that this possibility is central to class struggle. But we see the organic human capacity to *care* about things (good things or bad things) as the point of departure between humans and machines, and their consequent value generating capacities.

3. We do not speak of "digital capitalism," or prefix capitalism, or labor with a modifier, because it seems to us that it is in capitalism's very nature to constantly change, drawing from a deep well of flexibility and resilience. We can understand its changes through the development of concepts such as heteromation.

4. We are indebted to Professor Christian Wagner of the City University of Hong Kong, one of the first scholars to study Wikipedia (Wagner 2004), for sharing this knowledge about the uses of Wikipedia.

5. We are, therefore, not concerned with mainframe computers, enterprise resource planning systems, and the like. Our emphasis on individual persons makes our work complementary to that of Schiller (2014), who documents, in beautiful detail, policies and patterns of technology development and innovation at the organizational level.

6. http://www.lyricsmode.com/lyrics/c/country_joe_and_the_fish.

7. When Bonnie began working at Hewlett Packard in the 1990s, long-term employees recalled the "nine-day fortnight" of previous eras, in which employees temporarily received reduced pay for (sometimes) reduced work in an effort to decrease or eliminate layoffs during economic downturns. This policy reflected the Depression era values of Bill Hewlett and Dave Packard, or "Bill and Dave," as they continued to be known in the company during that era.

8. http://www.ebri.org/pdf/publications/facts/0205fact.a.pdf.

9. On February 19, 2015, headlines included "'Parks and Rec' Exec Harris Wittels Dies from Overdose" (he had been in rehab) and "How to Be a Charming Cheapskate in the Tinder era," as well as a story on *50 Shades of Grey*.

Chapter 4

1. Such career disappointments were chronicled in the TV show *Breaking Bad*, in which Walter White, a high school chemistry teacher who missed the chance to become a career scientist, deploys his knowledge of chemistry to get rich cooking high-grade methamphetamine. Though the show is based on a fanciful premise, Walt's bitterness is all too real, and resonated with millions of viewers. *Breaking Bad* is an artistic rendering of the predicament of futility. Many critics consider it the best TV show ever made, and it seems no accident that it digs deeply into contemporary pain, anger, and frustration.

2. Interest rates are typically higher at online "sharing economy" services than at banks (Karabell 2015).

3. See https://student.societyforscience.org/article/choosing-shocks-over-contemplation.

4. Futility and monotony contribute to increasing use of lethal drugs. In the United States, heroin deaths quadrupled between 2002 and 2013, according to the Centers for Disease Control and Prevention (CDC). People may start abusing prescription painkillers and then move to heroin, which is readily available. http://www.nbcnews.com/health/health-news/heroin-deaths-quadruple-across-us-n388006.

Chapter 5

1. See http://www.forbes.com/sites/samanthasharf/2015/02/05/linkedin-stock-pops-on-44-sales-growth.

2. See http://www.acxiom.com/partners.

3. See http://www.acxiom.com/about-acxiom/privacy/consumer-data-information.

4. Mark Zuckerberg famously said: "You have one identity. The days of you having a different image for your work friends or co-workers and for the other people you know are probably coming to an end pretty quickly. ... Having two identities for yourself is an example of a lack of integrity" (quoted in Kirkpatrick 2010, p. 199).

5. A 2011 survey showed that an absolute majority of employers screen potential employees through social networking sites such as Facebook and LinkedIn (Swallow 2011).

6. See www.acxiom.de/resources/audience-propensities.

Chapter 6

1. See http://www.npr.org/sections/alltechconsidered/2014/01/01/255739234/banks-try-to-save-big-with-atms-of-the-future.

2. See http://bankinnovation.net/2012/12/jpmorgan-chase-steps-up-installation-of-self-service-kiosks-and-card-issuing-machines-in-branches.

3. See http://electronics.howstuffworks.com/interactive-voice-response4.htm.

4. See http://www.interactions.net/news-items/research-reveals-consequences-of-companies-poor-automated-phone-systems.

5. See http://www-03.ibm.com/products/retail/products/self/sco/index.html.

6. Some players would like to see The Tribunal return, as evidenced in forum discussions (Yubo Kou, personal communication).

7. See http://www.datatrend.com/library/CreatingaSelf-ServiceWorld-0610.pdf.

8. We note that rich people typically choose institutions with excellent customer service and plenty of humans on offer, even fetishizing service in venues such as luxury safaris, trips to the summit of Everest, hotel rooms that come with butlers, etc. Capitalists who want to put kiosks in pharmacies undoubtedly have staff to pick up their prescriptions for them.

Chapter 7

1. One gaming website could not have put it more plainly: "The real joy of playing a game on a PC is that, thanks to mods, your entire experience can be improved by the work of dedicated fans" (http://kotaku.com/5979860/the-12-best-mods-for-pc-games).

2. One might think that mods would be of inferior quality. A modder we interviewed early in our research said that this is not at all true. He had been modding since the World of Warcraft beta release in 2004, and he told us that the modders he knew were "stinking smart" and produced high-quality software. Bonnie came to agree; she has used mods extensively in researching World of Warcraft. Many are exceptionally well designed. The quality is not surprising since players know what they want and design to their own needs, as against the gaps and missteps present in many commercial software systems, in which specifications guide development but lack the nuanced understanding practitioners can bring to the details. Bugs in mods are reported by players and quickly fixed by modders, who take pride in their creations.

3. Of course, there is never a level playing field. Kow and Nardi (2009, 2014) found institutional barriers to modding in China. Chinese modders did not have the same access to Blizzard employees (through forums and events such as BlizzCon) as American modders did. Chinese modders produced mods, but not nearly as many as American modders, despite the fact that there were more Chinese players of the game.

4. See https://www.youtube.com/watch?v=DsstOs-K7gk.

Chapter 8

1. We are concerned here with human labor, but it is worth noting that people appropriate the caring of other living creatures for economic purposes and have done so for centuries, even developing lines of animals (like dogs) through programs of selective breeding that produce extraordinarily caring animals.

2. It seems important to mention that such responses might also come from stuffed animals introduced by caregivers in the same manner PARO is introduced, i.e., with little stories and the encouragements of caregivers and others.

3. See the video *PARO for Patients in Italy* at the official PARO website (http://www.parorobots.com/video.asp). The elderly person's disinterest in PARO is evident, as well as the bleak conditions endured by both patients and caretakers in the nursing home. Eventually the caretaker inveigles the patient into interacting with PARO, which must have seemed a success to the video creators. But the caretaker's lonely job and the demented patient's inability to understand PARO are visible and poignant.

4. In Japan, with its graying population and resistance to immigration, robots will likely play an increasingly significant role in eldercare. Whether they will be successful as individual units, without much human assistance, remains to be seen. Remembering Harlow's monkeys and what they showed about attachment,

Notes 229

we expect that the elderly in heteromated situations will fare better than those who have little human interaction.

5. See http://www.thejournal.ie/help-the-aged-1814698-Dec2014.

Chapter 9

1. Services such as Uber are not heteromated in terms of drivers' labor, but through the labor of customers producing reviews. Customers are asked to rank and comment on their drivers. The company uses the reviews to discipline and manage drivers (see Raval and Dourish 2016).

2. See http://www.geo-wiki.org/oldgames/croplandcapture.

3. See http://publiclab.org/wiki/kite-mapping.

4. Even Wikipedia is not immune to capitalist appropriation of its content, as noted in chapter 4.

5. The metaphor of the "backbone" provides a useful window into the thinking but also into the reality of network arrangements. Behind the hype of network-enabled decentralization, lean organizations, and a flat world (e.g., Friedman 2005), what materializes most often is the displacement of layered pyramids by back-boned networks.

Chapter 10

1. As in any two-dimensional representation, we had to make choices about how to organize the cases, so the three logics are implicit. We chose to highlight the two types of participants since the shift from the normal distribution to the highly skewed "winners" and "losers" distribution of contemporary capitalism has been important to our argument, and is perhaps more in need of visualization than the categories of reward.

2. D. Poeter, "Norway Retailers Pull Violent Games in Wake of Mass Murder," PC, August 4, 2011, http://www.pcmag.com/article2/0,2817,2390479,00.asp; "Norway Agrees to Security Review After Killings," http://ca.reuters.com/article/topNews/idCAL6E7IN00C20110727.

3. Another example from Brynjolfsson and McAfee is Intuit's TurboTax. This program has certainly put some tax preparers out of business, but using it is hardly the "automated" process the authors say it is (TurboTax "automate[s] the job of tax preparation.") The taxpayer must do a lot of work to prepare her own taxes, spending hours figuring how to use the software, learning the tax codes, and so on. Often this work involves seeking help from friends and relatives, calling one's brokerage to determine, e.g., how to report foreign income, and so on. Tax preparation with TurboTax is an instance of one type of heteromation—computer-mediated labor is shifted to the end user, and a paid job is eliminated.

4. The empirical studies of modders, video creators, bloggers and others discussed in part II show that they do not initiate their activities with money in mind. PewDiePie, for example, has remarked that after he dropped out of college, he

was very happy selling hot dogs and making his videos—at the time he was neither rich nor famous.

5. There are indirect forms of financial gain for visible participants not shown in the table. For example, Twitter does not pay anyone to post, but celebrities may post endorsements paid for by others. For instance, Snoop Dogg endorsed a Toyota minivan on his Twitter feed in 2011 (see http://www.cbsnews.com/news/celebrities-paid-thousands-for-endorsement-tweets).

6. Celebrities have appropriated social media as extensions of their other "work of fame" (such as being interviewed for print media), and while this is important, our concerns go beyond social media. We are also interested in how the internet has opened up possibilities for ordinary people as a new way to derive value in networks (beyond what celebrities can do).

7. A related scheme, one emphasizing mobility, is presented in Ekbia (2016).

8. One microworker said, "There's no sick leave, paid holidays, anything like that on mTurk. There is no arbitration, no appeal if you feel that you have been unfairly treated, apart from a stinging review on Turkopticon [a university-supported worker forum]" (Hodson 2013).

9. Exceptions are hyper-alert users who deploy programs such as DuckDuckGo, Tor, and Bitcoin, as well as cash, to avoid (to the extent possible) data harvesting and tracking.

10. A job ad for the company reveals neoliberal underpinnings: "Health IQ was founded on the premise that people who take responsibility for their health are overpaying for a number of financial products because they are subsidizing those who are less Health Conscious." The company's vacation policy is "take what you need." No expensive HR personnel calculate how much time you receive down to the hour; employees choose for themselves. It is difficult not to applaud the following part of the explanation of employee benefits: "However, the real benefit is that unlike most startups, where you tend to gain 15 lbs (the Google 15) in the first few months after you join, our employees tend to lose 15 lbs in their first few months because there is no candy, soda, energy bars, etc. in our office. We have lots of nuts, fruits, vegetables, a squat rack, and a conference room made up of four walking treadmills that face each other." http://vput.io/vacancy/3431.

Chapter 11

1. The story is more convoluted in the United States, where the battle for universal health coverage is still going on. While Medicaid and Medicare, which began in the 1960s, cover some people, most people have required private insurance since the 1950s (when it became common for people to insure themselves—before that, most people had meager or no insurance).

2. During this era, African-American citizens had few opportunities to buy life insurance, which was sold exclusively through agents recruited through personal social networks. Residential segregation meant that few African-Amerians found work selling insurance, and their communities were therefore not offered the

Notes   231

product. Such discrimination persisted for decades, denying African-Americans an important form of protection.

3. See http://vput.io/vacancy/3431.

4. For a similar argument, presented as black humor, see Bob Black's *The Abolition of Work and Other Essays* from the 1980s (http://inspiracy.com/black/abolition/part1.html). Black is an attorney, pundit, humorist, and radical thinker.

Chapter 12

1. There is a vast literature on utopias that we will not engage here; see the papers in Bradley and Hedrén (*Green Utopianism* 2014) for recent treatment, and compelling empirical cases.

2. Various metrics reveal the extremity of the endgame and the income disparity it produces, for example, the revelation, in 2016, that 62 people control as much wealth as the bottom economic half of the world's population (http://money.cnn.com/2016/01/17/news/economy/oxfam-wealth/index.html).

3. Hornborg (2014, p. 92) remarks, "As Georgescu-Roegen realized, there are ... absolute physical limits to growth and resource extraction, ultimately defined by the laws of thermodynamics. The inclination of most economists and proponents of growth to dismiss this obvious ecological truth is remarkable."

4. Admittedly, Americans work longer hours than most people in industrialized nations. But it is difficult to imagine France and Northern Europe continuing, e.g., generous vacation benefits, with the millions of immigrants changing their economies. Things will shift as resources are increasingly reallocated to new demands and priorities begin to change. Austerity, or at least less abundant benefits, is the likely result, as has happened, for various reasons, elsewhere.

5. The reasons are extremely complicated, and we won't go into them here, but they materially concern capitalism's long history of global domination and exploitation.

6. It is difficult to say why DIY has recently struck a chord in computing research; perhaps people feel the dangers in our current system, and reach unconsciously for orientations that seem to suggest alternatives.

7. The history of the Bay Area Rapid Transit system is a case in point of the undesirability of too much local control. BART extends south from San Francisco only to the city of Millbrae. In 1962, the good citizens of San Mateo County, already a populated and prosperous region, decided they did not need rapid transit. However, now, south beyond Millbrae, lies Silicon Valley, arguably the most important economic hub in the world. Most of San Mateo County and all of Santa Clara County (i.e., Silicon Valley) have no rapid transit. Without track in San Mateo County, north of Santa Clara County, Santa Clara County gets no rapid transit. The costs and political hassles of installing infrastructure today are prohibitive. Just explaining these *local* governmental units and their need to cooperate is confusing! Traffic is increasingly congested in Silicon Valley, and the lack of

transportation alternatives impacts the poor most severely. It has given rise to controversial classist patchings-up such as the Google buses.

8. It seems unrealistic to expect that we will somehow transition to communism, as proponents of "cognitive capitalism" such as Vercellone (2007) hope.

9. As journalist Nathan Schneider points out, *how* a basic income is implemented is critical. There are proposals from the tech industry in which corporations would own and distribute everything (instead of the state). The safety net would be dismantled in favor of a social income derived from corporate profits (involving no doubt, a lot of heteromation). We are not in favor of such a scheme, which puts too much power into the hands of organizations whose goal is profit rather than the common weal. See http://americamagazine.org/issue/basic-justice.

10. The social income is not a panacea for homelessness, at least in the United States. A cohort of homeless are mentally ill or addicted to various substances, and need other services. The social income has variants such as the negative income tax which we do not go into here.

11. We might even borrow practices such as "pedestrian right" from Islamic culture that allow pedestrians or those passing by to eat from the fruits of trees or other plants to overcome hunger. Food cannot be removed from the premises or plants damaged (Emad Khazraee, personal communication).

12. In the second author's neighborhood, the local listserv is an efficient means of redistributing all kinds of things. A neighbor announces the items he or she will put outside their house, and the early bird gets the worm. Nearly all items are free although sometimes an expensive item may be priced. Once the item is claimed, the neighbor lets everyone know on the list. Cleaning out the garage has never been simpler.

13. At the time of this writing, the Seattle City Council passed a bill to allow drivers for Uber, Lyft, and such companies to unionize and negotiate collective agreements with the platform operating companies. The US Chamber of Commerce believes the Seattle legislation is illegal under federal antitrust law, and is fighting it.

14. Other approaches, such as schemes of local governance that require a good deal of organization and politicking, are discussed by Wright (2009).

15. There is a deep history here going back to World War I (see Hayden-Smith 2014).

16. Such programs include DuckDuckGo for search, BitTorrent Sync for file sharing, Meteor for prototyping web applications, HotCRP for conference management, and a host of other free systems.

17. Hofstadter (1979) coined the term quine, a theoretical concept that now refers to self-reproducing software systems.

18. Rifkin (2014) says that we will manufacture things on our 3D printers, but we will buy the printers at Staples. The value chain here is hardly a simple one of "free" goods from the Commons. The book is filled with numerous similar examples; e.g., architects and engineers are "scrambling" to bring 3D printed buildings to market (Rifkin 2014). The Commons shifts in and out of view as the central

motif of the zero marginal cost society. Rifkin's cases often concern conventional profit-making opportunities carried out in conventional enterprises (such as 3D printed cars, and so on.).

19. Rifkin's proposals for energy cooperatives seem sound; he notes that 42 million customers in the United States are served by non-profit rural electric cooperatives. Currently, unfortunately, most state utilities are organized on a business model, as profit-making entities. We hope the rural cooperative model can become more widespread. Since its inception in the 1930s, cooperatives have worked out how to run utilities without profits and they are very successful.

20. See http://dark-mountain.net. We are critical of Dark Mountain, but suggest reading their books and blogs, which are thoughtful, and often beautifully written.

Epilogue

1. One of the ironies of counseling education as the panacea to current social problems is that many of the big winners of the new economy are college dropouts, including characters as disparate as Steve Jobs and PewDiePie.

2. Felix Salmon, "Krugman vs. Summers: The Debate." Reuters. November 15, 2011, accessed January 12, 2016, fromhttp://blogsreuters.com/felix-salmon/2011/11/15/krugman-vs-summers-thedebate.

3. Far from being a fictitious person, this is an accurate description of a close relative that Hamid knows and cares about immensely. If we look around carefully, all of us can easily identify educated people who work for corporations with deceptive titles such as "manager," only to be deprived of overtime payments, weekends, sick leave, and even holidays.

4. In fairness, they also provide some technical economic arguments to support their theory of "technological unemployment." We don't need to get into these arguments here because we are trying to make a more basic point on economistic thinking of human beings.

5. The canonical story to illustrate this shortcoming of the human mind is the fiction of the Chinese emperor who was misled by the inventor of chess to reward him by doubling, each day, the number of grains on each square of the board—a number that became unimaginably large when it got to the second half of the board. In our times, the hardware engine that drives exponential growth is allegedly guaranteed by Moore's Law—an engineering proposal that computing speed doubles every 18 months. The success of the computer industry in upholding this law for many decades has incited various commentators to make far-fetched and, in our view, unfounded predictions about the relationship of humans and machines—e.g., that computers will soon surpass us in "intelligence" (see Ekbia 2008).

References

Alač, M., J. Movellan, and F. Tanaka. 2011. When a Robot Is Social: Spatial Arrangements and Multimodal Semiotic Engagement in the Practice of Social Robotics. *Social Studies of Science* 41 (6): 893–926.

Alaimo, C., and J. Kallinikos. 2016. Encoding the Everyday: The Infrastructural Apparatus of Social Data. In *Big Data Is Not a Monolith*, ed. C. Sugimoto, H. Ekbia and M. Mattioli, 77–90. Cambridge, MA: MIT Press.

Altieri, M. 1999. The Ecological Role of Biodiversity in Agroecosystems. *Agriculture, Ecosystems & Environment* 74:19–31.

André, P., H. Zhang, J. Kim, L. Chilton, S. Dow, and R. Miller. 2013. Community Clustering: Leveraging an Academic Crowd to Form Coherent Conference Sessions. *Proceedings HCOMP '13,* Palm Springs.

Andrejevic, M. 2013. Estranged Free Labor. In *Digital Labor: The Internet as Playground and Factory*, ed. Trebor Scholz, 149–165. New York: Routledge.

Andrejevic, M. 2015. Personal Data: Blind Spot of the "Affective Law of Value"? *The Information Society* 31 (1).

Anglietta, M. 1979. *A Theory of Capitalist Regulation*. New Left Books.

Arendt, H. 1958. *The Human Condition*. Chicago: The University of Chicago Press.

Arrow, K. 1969. The Organization of Economic Activity: Issues Pertinent to the Choice of Market versus Nonmarket Allocation. In *The Analysis and Evaluation of Public Expenditure: The PPB System*. Vol. 1, US Joint Economic Committee, 91st Congress, 1st Session, Washington DC: US Government Printing Office, pp. 59–73.

Arvidsson, A., and E. Colleoni. 2012. Value in informational capitalism and on the Internet. *Information Society* 28 (3): 135–150.

Ashby, R. 1956. *An Introduction to Cybernetics*. London: Chapman and Hall.

Assange, J. 2014. *When Google Met WikiLeaks*. New York: OR Books.

Auger, J. 2014. Living with Robots: A Speculative Design Approach. *Journal of Human—Robot Interaction* 3 (1): 20–42.

Autor, D. 2014. Skills, Education, and the Rise of Earnings Inequality among the "Other 99 Percent." *Science* 344 (6186): 843–851.

Babbage, C. 1832. *On Machinery and Manufactures*. London: Charles Knight.

Bailey, D., E. Diniz, and D. Scholler. 2014. Achieving ICT4D Project Success by Altering Context, Not Technology. *ICTD* 10 (4): 15–29.

Bailey, D., E. Diniz, B. Nardi, P. Leonardi, and D. Scholler. 2016. Multiplex Appropriation in Complex Systems Implementation: The Case of Brazil's Correspondent Banking System. *Management Information Systems Quarterly*.

Barabási, A. L. 2002. *Linked: The New Science of Networks*. New York: Perseus Books.

Baumer, E., P. Adams, V. Khovanskaya, T. Liao, M. Smith, V. Sosik, and K. Williams. 2013. Limiting, Leaving, and (re)Lapsing: An Exploration of Facebook Non-Use Practices and Experiences. *Proceedings of the ACM Conference on Human Factors in Computer Systems*. Pp. 3257–3266.

Baumer, E., and M. Silberman. 2011. When the Implication Is Not to Design (Technology). *Proceedings of the ACM Conference on Human Factors in Computer Systems*, 2271–2274.

Beck, U. 1992. *Risk Society: Towards a New Modernity*. London: Sage.

Becker, G. 1962. Irrational Action and Economic Theory. *Journal of Political Economy* 70 (1): 153–168.

Becker, G. 1965. *The Economic Approach to Human Behavior*. Chicago: Chicago University Press.

Becker, H. 1960. Notes on the Concept of Commitment. *American Journal of Sociology* 66 (1): 32–40.

Becker, H. 2004. Interaction: Some Ideas. Paper presented at the Université Pierre Mendes-France, Grenoble. http://howardsbecker.com/articles/interaction.html.

Beer, D. 2009. Power through the Algorithm? Participatory Web Cultures and the Technological Unconscious. *New Media & Society* 11 (6): 985–1002.

Bejerot, N. 1974. The Six Day War in Stockholm. *New Scientist* 61 (886): 486–487.

Bell, D. 1976. *The Coming of Post-Industrial Society: A Venture in Social Forecasting*. Harmondsworth: Penguin, Peregrine Books.

Bellotti, V., S. Cambridge, K. Hoy, P. Shih, L. Handalian, K. Han, and J. Carroll. 2014. *Towards Community-Centered Support for Peer-to-Peer Service Exchange: Rethinking the Timebanking Metaphor*, 2975–2984. Proceedings NordiCHI.

Beniger, J. 1986. *The Control Revolution*. Cambridge: Harvard University Press.

Benkler, Y. 2002. Coase's Penguin, or, Linux and The Nature of the Firm. *Yale Law Journal* 112:369–446.

Benkler, Y. 2006. *The Wealth of Networks: How Social Production Transforms Markets and Freedom*. New Haven: Yale University Press.

Berardi, F. 2009. *The Soul at Work: From Alienation to Autonomy*. Cambridge, MA: MIT Press.

Blevis, Eli. 2007. Sustainable Interaction Design: Invention & Disposal, Renewal & Reuse. *Proceedings of the ACM Conference on Human Factors in Computer Systems*. Pp. 503–512.

Boltanski, L., and È. Chiapello. 2005. *The New Spirit of Capitalism*. New York: Verso.

Boltanski, L., and L. Thévenot. 2006. *On Justification. The Economies of Worth*. Princeton, NJ: Princeton University Press.

Booth, D. 1998. *The Environmental Consequences of Growth: Steady-state Economics as an Alternative to Ecological Decline*. London: Routledge.

Bradley, K. 2014. Towards a Peer Economy. In *Green Utopianism*, ed. K. Bradley and J. Hedrén, 183–204. New York: Routledge.

Broekens, J., M. Heerink, and H. Rosendal. 2009. Assistive Social Robots in Elderly Care: A Review. *Gerontechnology (Valkenswaard)* 8 (2): 94–103.

Brooks, R. 1999. *Cambrian Intelligence: The Early History of the New AI*. Cambridge, MA: MIT Press.

Brown, W. 2015. *Undoing the Demos*. Cambridge, MA: MIT Press.

Bruder, J. 2014. The End of Retirement: When You Can't Afford to Stop Working. *Harper's Magazine*, August. Pp. 28–36.

Brynjarsdóttir, H., M. Håkansson, J. Pierce, E. Baumer, C. DiSalvo, and P. Sengers. 2012. Sustainably Unpersuaded: How Persuasion Narrows Our Vision of Sustainability. *Proceedings of the ACM Conference on Human Factors in Computer Systems*. Pp. 947–956.

Brynjolfsson, E., and A. McAfee. 2011. *Race Against the Machine*. Digital Frontier Press.

Brynjolfsson, E., and A. McAfee. 2016. *The Second Machine Age*. New York: W.W. Norton and Co.

Buchanan, J. 1999. *The Logical Foundations of Constitutional Liberty*. New York: Liberty Fund.

Buchsbaum, J. 2016. The Exceptional *Intermittents du Spectacle*. In *The Routledge Companion to Labor and Media*, ed. R. Maxwell, 154–169. New York: Routledge.

Bullard, J. 2013. It Takes a Jerk to Make a Conversation into an Archive. *Proceedings iConference*. Pp. 583–588.

Burawoy, M. 1979. *Manufacturing Consent: Changes in the Labor Process Under Monopoly Capitalism*. Chicago: University of Chicago Press.

Burdon, M., and M. Andrejevic. 2016. Big Data in the Sensor Society. In *Big Data Is Not a Monolith*, ed. C. Sugimoto, H. Ekbia and M. Mattioli, 51–75. Cambridge, MA: MIT Press.

Burkett, P. 2003. The Value Problem in Ecological Economics: A Marxist Intervention. *Historical Materialism* 13 (1): 117–152.

Bush, V. 1945. As We May Think. *The Atlantic Monthly*. http://www.ps.unisaarland.de/~duchier/pub/vbush/vbush.txt

Caffentzis, G. 2005. Immeasurable Value? An Essay on Marx's Legacy. *The Commoner* (10): 87–114.

Caffentzis, G. 2013. *In Letters of Blood and Fire*. Oakland: PM Press.

Callois, R. 1961. *Man, Play, and Games*. New York: Free Press.

Campbell-Kelly, M., and W. Aspray. 1996. *Computer: A History of the Information Machine.* New York: Basic Books.

Campbell-Kelly, M., W. Aspray, N. Ensmenger, and J. Yost. 2013. *Computer: A History of the Information Machine.* New York: Westview Press.

Castells, M. 1989. *The Rise of the Network Society.* Malden, MA: Blackwell.

Castells, M. 2009. *Communication Power.* Oxford: Oxford University Press.

Cerratto-Pargman, T., D. Pargman, and B. Nardi. 2016. The Internet at the Eco-Village: Performing Sustainability in the Twenty-First Century. *First Monday* 21 (5), May 2. http://firstmonday.org/ojs/index.php/fm/article/view/6637.

Cervantes, R., M. Warschauer, B. Nardi, and S. Sambasivan. 2011. Infrastructures for Low-Cost Laptop Use in Mexican Schools. *Proceedings of the ACM Conference on Human Factors in Computer Systems.* Pp. 945–954.

Chamayou, G. 2014. *A Theory of the Drone.* New York: The New Press.

Chandler, A. 1980. The United States: Seedbed of Managerial Capitalism. In *The Transformation of Industrial Organization: Management, Labor, and Society in the United States*, ed. F. Hearn, 34–45. New York: Wadsworth.

Chang, W.-L., S. Šabanović, and L. Huber. 2013. Situated Analysis of Interactions between Cognitively Impaired Older Adults and the Therapeutic Robot PARO. *Proceedings ICSR* 2013:371–380.

Chang, W.-L., S. Šabanović, and L. Huber. 2014. Observational Study of Naturalistic Interactions with the Socially Assistive Robot PARO in a Nursing Home. *Proceedings of RO-MAN' 2014.*

Chilton, L., J. Horton, R. Miller, and S. Azenkot. 2010. Task Search in a Human Computation Market. *Proceedings KDD-HCOMP.*

Choontanom, T., and B. Nardi. 2012. Theorycrafting: The Art and Science of Using Numbers to Interpret the World. In *Games, Learning, and Society: Learning and Meaning in the Digital Age*, ed. C. Steinkuehler, K. Squire and S. Barab. Cambridge: Cambridge University Press.

Coase, R. 1960. The Problem of Social Cost. *Journal of Law & Economics* 3:1–44.

Cohen, S. 1973. *Folk Devils and Moral Panics.* London: St. Albans.

Comor, E. 2015. Revisiting Marx's Value Theory: A Critical Response to Analyses of Digital Prosumption. *Information Society* 31 (1): 13–19.

Conaty, P., and D. Bollier. 2014. Toward an Open-Cooperativism: A New Social Economy Based on Open Platforms, Co-operative Models and the Commons. A Report on a Commons Strategies Group Workshop. Berlin, Germany August 27–28.

Cornwall, A. 2008. Unpacking "Participation": Models, Meanings and Practices. *Community Development Journal* 43 (3): 269–283.

Croll, A. 2013. Big Data Is Our Generation's Civil Rights Issue, and We Don't Know It. http://radar.oreilly.com/2012/08/big-data-is-our-generations-civil-rights-issue-and-we-dont-know-it.html.

CUSLAR. 2014. The Hardships and Dehumanizing Struggles of a Globalized Immigrant Population. http://cuslar.org/2014/03/06/the-hardships-and-dehumanizing-struggles-of-a-globalized-immigrant-population.

Daft, R. 2001. *Organization Theory and Design*. 7th ed. Cincinnati: South-Western College Publishing.

Daly, H. 1991. *Steady State Economics*. Washington, DC: Island Press.

De Brunhoff, S. 2005. Marx's Contribution to the Search for a Theory of Money. In *Marx's Theory of Money: Modern Appraisals*, ed. F. Mosele, 209–221. London: Palgrave Macmillan.

Deleuze, G. 1992. *Postscript on the Societies of Control. Originally published in the journal OCTOBER 59, Winter 1992*, 3–7. Cambridge, MA: MIT Press.

Dertouzos, M. 1979. Individualized Automation. *The Computer Age: A Twenty-Year View*, eds. M. Dertouzos and J. Moses, 38–55. Cambridge, MA: MIT Press.

De Stefano, V. 2016. The Rise of the "Just-In-Time Workforce": On-Demand Work, Crowdwork and Labour Protection in the "Gig-Economy." *Comparative Labor Law & Policy Journal* 37 (3).

Dewey, J. 2005. *Art as Experience*. New York: Perigee. (First published 1934.)

Dillahunt, T. 2014) Toward a Deeper Understanding of Sustainability within HCI. *Proceedings of the ACM Conference on Human Factors in Computer Systems, Extended Abstracts*.

Diniz, E. 2007. Correspondent Banking and Microcredit in Brazil: Banking Technology and the Expansion of Financial Services for the Low Income Population. São Paulo: GVPesquisa Report.

Diniz, E., D. Bailey, S. Dailey, and D. Sholler. 2013. Bridging the ICT4D Design-Actuality Gap: "Human ATMs" and the Provision of Financial Services for "Humble People." *International Conference on Information Resources Management* 2013. Natal, Brazil.

Dunne, A., and F. Raby. 2013. *Speculative Everything: Design, Fiction, and Social Dreaming*. Cambridge: MIT Press.

Dyer-Witheford, N. 1999. *Cyber-Marx*. Urbana: University of Illinois Press.

Employee Benefit Research Institute. 2005. History of 401(k) Plans: An Update. http://www.ebri.org/pdf/publications/facts/0205fact.a.pdf.

Ekbia, H. R. 2008. *Artificial Dreams: The Quest for Non-Biological Intelligence*. New York: Cambridge University Press.

Ekbia, H. R. 2009. Digital Artifacts as Quasi-Objects: Qualification, Mediation, and Materiality. *Journal of the American Society for Information Science and Technology* 60 (12): 2554–2566.

Ekbia, H. R. 2012. Heteronomous Humans and Autonomous Agents: Toward Artificial Relational Intelligence. In *Beyond Artificial Intelligence: The Disappearing Human-Machine Divide: Topics in Intelligent Engineering and Informatics*. vol. 9. Ed. J. Fodor and I. Ruda, 63–78. Heidelberg: Springer.

Ekbia, H. R. 2016. Digital Inclusion and Social Exclusion: The Political Economy of Value in a Networked World. *Information Society*.

Ekbia, H. R., and B. Nardi. 2012. Inverse Instrumentality: How Technologies Objectify Patients and Players. In *Materiality and Organizing: Social Interaction in a Technological World*, ed. P. Leonardi, B. Nardi and J. Kallinikos, 157–176. Oxford: Oxford University Press.

Ekbia, H. R. and Nardi, B. 2014. Heteromation and Its (Dis)contents: The Invisible Division of Labor between Humans and Machines. *First Monday*, June.

Ekbia, H. R., and B. Nardi. 2015. *The Political Economy of Computing: The Elephant in the HCI Room. Interactions*, 46–49. New York: ACM Press.

Ekbia, H. R., and B. Nardi. 2016. Social Inequality and HCI: The View from Political Economy. *Proceedings of the ACM Conference on Human Factors in Computer Systems*. New York: ACM Press.

Ekbia, H. R., Nardi, B. and Šabanović, S. 2015. On the Margins of the Machine: Heteromation and Robotics. *Proceedings iConference*.

Elgin, D., and A. Mitchell. 1977. Voluntary Simplicity. *The Co-Evolution Quarterly*, Summer. http://www.duaneelgin.com/wp-content/uploads/2010/11/voluntary_simplicity.pdf.

Emirbayer, M., and A. Mische. 1998. What Is Agency? *American Journal of Sociology* 103:962–1023.

Engelbart, D. 1962. Augmenting Human Intellect: A Conceptual Framework. Research Report. DougEngelbart.org.

Etzioni, A. 1988. *The Moral Dimension: Toward a New Economics*. New York: The Free Press.

Facebook. 2012. Registration Statement. United States Securities And Exchange Commission, Washington, D.C. https://www.sec.gov/Archives/edgar/data/1326801/000119312512034517/d287954ds1.htm.

Federici, S. 2012. Wages against Housework. In R*evolution at Point Zero: Housework, Reproduction, and Feminist Struggle*. Oakland: PM Press/Common Notions. (Originally published in *The Politics of Housework*, E. Malos, ed. London: Allison and Busby, 1982.)

Ferguson, R. A. 2013. *Alone in America: The Stories that Matter*. Cambridge, MA: Harvard University Press.

Ferguson, R. S., and S. Lovell. 2014. Permaculture for Agroecology: Design, Movement, Practice, and Worldview: A Review. *Agronomy for Sustainable Development* 34 (2): 251–274.

Fisher, D., and M. Fisher. 1996. *Tube: The Invention of Television*. Washington, DC: Counterpoint.

Forbes. 2013. The World's Billionaires List. www.forbes.com/billionaires/list.

Forlizzi, J., and C. DiSalvo. 2006. Service Robots in the Domestic Environment: A Study of the Roomba VACUUM in the home. *Proceedings of the ACM SIGCHI/SIGART Conference on Human-RobotInteraction (HRI)*.

Forte, A., V. Larco, and A. Bruckman. 2009. Decentralization in Wikipedia Governance. *Journal of Management Information Systems* 26 (1): 49–72.

Foster, J. 2002. *Ecology Against Capitalism*. New York: Monthly Review Press.

Foucault, M. 1977. *Discipline and Punish*. Trans. R. Howard. New York: Vintage Books.

Foucault, M. 1986. Space, Knowledge, and Power. In *The Foucault Reader*, ed. P. Rabinow. Harmondsworth: Penguin.

Foucault, M. 1991. Why This Prison? In *The Foucault Effect: Studies in Governmentality*, ed. G. Burchell, C. Gordon and P. Miller. Chicago: University of Chicago Press.

Foucault, M. 1997. *Ethics: Subjectivity and Truth*, ed. Paul Rabinow. New York: The New Press.

Foucault, M. 2008. *The Government of Self and Others: Lectures at the College de France 1982–83*. New York: Macmillan.

Fox, J. 2015. Rice Gets a Bath Amid California's Drought. *Bloomberg View*, May 15. http://www.bloombergview.com/articles/2015-05-15/california-floods-fields-to-grow-rice-in-a-drought.

Franklin, B. (1748). *Advice to a Young Tradesman, Written by an Old One*.

Freire, P. 2002. *Pedagogy of the Oppressed*. New York: Continuum. (First published 1970.)

Friedman, M. 1953. *Essays in Positive Economics*. Chicago: University of Chicago Press.

Friedman, M. 1962. *Capitalism and Freedom*. Chicago: University of Chicago Press.

Friedman, T. 2005. *The World Is Flat: A Brief History of the Twenty-First Century*. New York: Farrar, Straus and Giroux.

Fuchs, C. 2011. An Alternative View of Privacy on Facebook. *Information* 2: 140–165.

Fuchs. C. 2012. With or without Marx? With or without capitalism? A rejoinder to Adam Arvidsson and Eleanor Colleoni. *TripleC* 10 (2):633–645.

Fuchs, C., and S. Sevignani. 2013. What Is Digital Labour? What Is Digital Work? What's Their Difference? And Why Do These Questions Matter for Understanding Social Media? *tripleC: Communication, Capitalism & Critique. Open Access Journal for a Global Sustainable Information Society* 1:237–293.

Galbraith, J. K. 1985. *The Economics of Innocent Fraud: Truth for Our Time*. New York: Allen Lane.

Gassée, J.-L. 1987. *The Third Apple: Personal Computers and the Cultural Revolution*. New York: Harcourt.

Gates, B. 2007. A Robot in Every Home. *Scientific American* (January): 58–65.

Georgescu-Roegen, N. 1971. *The Entropy Law and the Economic Process*. Cambridge, MA: Harvard University Press.

Giddens, A. 1990. *The Consequences of Modernity*. Cambridge: Polity.

Gizmodo 2013. Google Finally Puts CAPTCHAS to Good Use. http://gizmodo.com/5897661/google-finally-puts-captchas-to-good-use.

Golding, D. 2015. The End of Gamers. In *State of Play*, ed. D. Goldberg and L. Larsson, 127–140. New York: Seven Stories.

Gordon, R. 2012. Is U.S. Economic Growth Over? Faltering Innovation Confronts the Six Headwinds. NBER Working Paper 18315. http://www.nber.org/papers/w18315.

Gorz, A. 1985. *Paths to Paradise*. New York: South End Press.

Gramsci, A. 1971. *Selections from the Prison Notebooks*. New York: International Publishers.

Greenspan, A. 1996. The Challenge of Central Banking in a Democratic Society, December 5. http://www.federalreserve.gov/boarddocs/speeches/1996/19961205.htm.

Grudin, J. 2012. A Moving Target: The Evolution of HCI. In *Human-computer Interaction Handbook: Fundamentals, Evolving Technologies, and Emerging Applications*. 3rd ed., ed. J. Jacko. London: Taylor & Francis.

Gui, X., and B. Nardi. 2015. Sustainability Begins in the Street: A Story of Transition Town Totnes. *Proceedings ICT4S*.

Håkansson, M., and P. Sengers. 2013. Beyond Being Green: Simple Living Families and ICT. *Proceedings of the ACM Conference on Human Factors in Computer Systems*. Pp. 2725–2734.

Harvey, D. 1999. *The Limits to Capital*. London: Verso.

Harvey, D. 2010. *The Enigma of Capital*. Oxford: Oxford University Press.

Haushofer, J., and E. Fehr. 2014. *Science* 344 (6186): 862–867. doi: 10.1126/science.1232491.

Hayden-Smith, R. 2014. *Sowing the Seeds of Victory: American Gardening Programs of World War 1*. Jefferson, North Carolina: McFarland.

Heidegger, M. 1962. *Being and Time*. New York: Harper and Row. (First published 1927.)

Heidegger, M. 1977. *The Question Concerning Technology and Other Essays*. New York: Harper and Row. (First published 1954.)

Hellman, M., T. Schoenmakers, B. Nordstrom, and R. van Holst. 2013. Is There Such a Thing as Online Video Game Addiction? A Cross-Disciplinary Review. *Addiction Research and Theory* 21 (2): 102–112.

Hobsbawm, E. 1999. *Industry and Empire: The Birth of the Industrial Revolution*. New York: New Press. (First published 1968.)

Hodson, H. 2013. Crowdsourcing Grows Up as Online Workers Unite. *New Scientist*. https://www.newscientist.com/article/mg21729036-200-crowdsourcing-grows-up-as-online-workers-unite.

Hofstadter, D. 1979. *Gödel, Escher, Bach: An Eternal Golden Braid*. New York: Basic Books.

Holmes, B. 2002. The Flexible Personality: For a New Cultural Critique. European Institute for Progressive Cultural Policies. http://eipcp.net/transversal/1106/holmes/en.

Holmes, B. 2005. Interview. *Journal of Aesthetics & Protest*. http://asounder.org/resources/holmes_flexiblepersonality.pdf.

Hornborg, A. 2001. *The Power of the Machine*. Walnut Creek, CA: Altamira Press.

Hornborg, A. 2014. Why Solar Panels Don't Grow on Trees. In *Green Utopianism*, ed. K. Bradley and J. Hedrén, 76–97. New York: Routledge.

Horton, J., and L. Chilton. 2010. The Labor Economics of Paid Crowdsourcing. Paper presented at the ACM Conference on Electronic Commerce, Boston, MA.

Hospitality Industry. 2009. Self-Service: The Next Generation. Self-Service Technology Study.

Hospitality Industry. 2011. Self-Service Tech Study.

Howard, J. 1969. The Flowering of the Hippie Movement. *Protest in the Sixties* (March): 43–55. Vol. 382 of *Annals of the American Academy of Political and Social Science*.

Huff, D. 1954. *How to Lie with Statistics*. New York: Norton.

Huffman, M. 2014. From Healthcare to Hamburgers, It's a Self-Service World. http://www.consumeraffairs.com/news/from-healthcare-to-hamburgers-its-a-self-service-world-100114.html.

Huizinga, J. 1950. *Homo Ludens: A Study of the Play Element in Culture*. Boston: Beacon Press.

Huntemann, N. 2014. No More Excuses: Using Twitter to Challenge The Symbolic Annihilation of Women in Games. *Feminist Media Studies* 15 (1): 164–167.

Hüttenraunch, H., A. Green, M. Norman, L. Oestreicher, and K. Severinson Eklundh. 2004. Involving Users in the Design of a Mobile Office Robot. *IEEE Transactions on Systems, Man and Cybernetics. Part C, Applications and Reviews* 34 (2): 113–124.

Huws, U. 2003. *The Making of a Cyberteriat: Virtual Work in a Real World*. New York: Monthly Review Press.

Inoue, K., K. Wada, and Y. Ito. 2008. Effective Application of Paro: Seal Type Robots for Disabled People in According to Ideas of Occupational Therapists. Paper presented at the Computers Helping People with Special Needs Conference.

IPCC. Intergovernmental Panel on Climate Change. 2014. *Climate Change 2013: The Physical Science Basis*. Working Group I Contribution to the Fifth Assessment Report of the Intergovernmental Panel on Climate Change, eds. T. Stocker, D. Qin, G.-K. Plattner, M. Tignor, S. Allen, J. Boschung, A. Nauels, Y. Xia, V. Bex, and P. Midgley. Cambridge: Cambridge University Press.

Irani, L. 2013. The Cultural Work of Microwork. *New Media & Society* 17 (5): 720–739.

Irani, L., and S. Silberman. 2013. Turkopticon: Interrupting Worker Invisibility in Amazon Mechanical Turk. *Proceedings of the ACM Conference on Human Factors in Computer Systems.* Pp. 16–21.

Jackson, S. 2014. Rethinking Repair. In *Media Technologies*, ed. T. Gillespie, et al. Cambridge, MA: MIT Press.

Jameson, F. 2011. *Representing Capital: A Reading of Volume One.* New York: Verso.

Jamieson, K., and J. Cappella. 2008. *Echo Chamber: Rush Limbaugh and the Conservative Media Establishment.* Oxford: Oxford University Press.

Jenkins, H. 2006. *Fans, Bloggers, and Gamers: Media Consumers in the Digital Age.* New York: NYU Press.

Jiang, L., C. Wagner, and B. Nardi. 2015. Not Just in It for the Money: A Qualitative Investigation of Workers' Perceived Benefits of Micro-task Crowdsourcing. *Proceedings HICSS*:773–782.

John, N. A. 2012. The Social Logics of Sharing: Web 2.0, Sharing Economies and the Therapeutic Narrative. In *Cultures and Ethics of Sharing/Kulturen und Ethiken des Teilens*, eds. T. Hug, R. Maier, and F. Stalder. Innsbruck: Innsbruck University Press.

Joshi, S., and T. Cerrato Pargman. 2015. On Fairness & Sustainability: Motivating Change in the Networked Society. *Proceedings ICT4S.*

Kaldor, N. 1962. *Capital Accumulation and Economic Growth. United Nations Educational Scientific and Cultural Organization. Seminar on the Programming of Economic Development.* Sao Paolo.

Kallinikos, J. 2004. Farewell to Constructivism: Technology and Context-Embedded Action. In *The Social Study of Information and Communication Technology*, ed. C. Avgerou, C. Ciborra and F. Land, 140–161. Oxford: Oxford University Press.

Kallinikos, J. 2011. *Governing through Technology: Information Artefacts and Social Practice.* London: Palgrave Macmillan.

Kallinikos, J., A. Aaltonen, and A. Marton. 2013. The Ambivalent Ontology of Digital Artifacts. *MIS Quarterly*, 37 (2): 357–370.

Kanda, T., R. Sato, N. Saiwaki, and H. Ishiguro. 2007. A Two-Month Field Trial in an Elementary School for Long-Term Human-Robot Interaction. *IEEE Transactions on Robotics* 23 (5): 962–971.

Kanda, T., M. Shiomi, Z. Miyashita, H. Ishiguro, and N. Hagita. 2009. An Affective Guide Robot in a Shopping Mall. *Proceedings of the ACM/IEEE International Conference on Human-Robot Interaction (HRI).*

Kaptelinin, V. 2014. Crafting User Experience of Self-Service Technologies: Key Challenges and Potential Solutions. *Proceedings DIS* (Design of Interactive Systems). Workshop paper.

Kaptelinin, V., and B. Nardi. 2006. *Acting with Technology: Activity Theory and Interaction Design.* Cambridge, MA: MIT Press.

Karabell, Z. 2015. The Uberization of Money. *Wall Street Journal.* http://www.wsj.com/articles/the-uberization-of-finance-1446835102.

Kessler, S. 2015. Why a New Generation of On-Demand Businesses Rejected the Uber Model. *Fast Company*. http://www.fastcompany.com/3058299/why-a-new-generation-of-on-demand-businesses-rejected-the-uber-model.

Kiesler, S., J. Siegel, and T. McGuire. 1984. Social Psychological Aspects of Computer-Mediated Communication. *American Psychologist* 39:1123–1134.

King, J. L., S. Iacono, and J. Grudin. 2004. Going Critical: Perspective and Proportion in the Epistemology of Rob Kling. *The Information Society*, 23 (4): 251–262.

Kirkpatrick, D. 2010. *The Facebook Effect: The Inside Story of the Company that is Connecting the World*. New York: Simon and Schuster.

Kirman, B., S. Lawson, J. Linehan, and D. O'Hara. 2013. CHI and the Future Robot Enslavement of Humankind: A Retrospective. *Proceedings of the ACM Conference on Human Factors in Computer Systems*. Pp. 2199–2208.

Klein, N. 2014. *This Changes Everything: Capitalism vs. the Climate*. New York: Simon and Schuster.

Kling, R., ed. 1996. *Computerization and Controversy: Value Conflicts and Social Choices*. San Diego: Academic Press.

Knoblauch, W. 2015. Game Over?: A Cold War Kid Reflects on Apocalyptic Video Games. In *State of Play*, ed. D. Goldberg and L. Larsson, 188–210. New York: Seven Stories.

Knowles, B., L. Blair, M. Hazas, and S. Walker. 2013. Exploring Sustainability Research in Computing. *Proceedings Ubicomp* 13:305–314.

Knowles, B., M. Lochrie, P. Coulton, and J. Whittle. 2014. Barter: A Technology Strategy for Local Wealth Generation. *IT Professional* 16 (3): 28–34.

Körner, A., M. Reitzle, and R. Silbereisen. 2012. Work-Related Demands and Life Satisfaction. *Journal of Vocational Behavior* 80:187–196.

Kou, Y., and B. Nardi. 2013. Regulating Anti-Social Behavior on the Internet: The Example of League of Legends. *Proceedings of the iConference*. Pp. 616–622.

Kou, Y., and B. Nardi. 2014. Governance in League of Legends: A Hybrid System. *Proceedings Foundations of Digital Games*.

Kow, Y., and B. Nardi. 2009. Culture and Creativity: World of Warcraft Modding in China and the U.S. In *Online Worlds: Convergence of the Real and the Virtual*, ed. W. S. Bainbridge. Heidelberg: Springer.

Kow, Y. M., and B. Nardi eds. 2010a. User Creativity, Governance and the New Media. *First Monday*, May.

Kow, Y. M. and Nardi, B. 2010b. Who Owns the Mods? *First Monday*, May.

Kow, Y. M., and B. Nardi. 2012. Mediating Contradictions of Digital Media. *UCI Law Review* 2:675–692.

Kow, Y. M., and B. Nardi. 2014. Rethinking Participatory Culture: Lessons from Core Teams in China. In *Videogames and Virtual Realities in East Asia*, ed. D. Wong and W. Kelly. London: Routledge.

Kücklich, J. 2005. Precarious Playbour: Modders and the Digital Games Industry. *Fibreculture* 5.

Kunda, G., S. Barley, and J. Evans. 2002. Why Do Contractors Contract? The Experience of Highly Skilled Technical Professionals in a Contingent Labor Market. *Industrial & Labor Relations Review* 55:234–261.

Kumar, A., A. Nair, A. Parsons, and E. Urdapilleta. 2006. Expanding Bank Outreach through Retail Partnerships: Correspondent Banking In Brazil. Washington: World Bank Working Paper No. 85.

Kurzweil, R. 1999. *The Age of Spiritual Machines*. New York: Viking.

Laney, D. 2012. To Facebook, You're Worth $80.95. CIO Journal: Wall Street Journal Blogs, May 3. http://blogs.wsj.com/cio/2012/05/03/to-facebook-youre-worth-80-95.

Langlois, R. 2003. Cognitive Comparative Advantage and the Organization of Work: Lessons from Herbert Simon's Vision of the Future. *Journal of Economic Psychology* 24:167–187.

Langlois, R. 2007. *The Dynamics of Industrial Capitalism: Schumpeter, Chandler, and the New Economy*. London: Routledge.

Lash, S., and J. Urry. 1987. *The End of Organized Capitalism*. Cambridge: Polity.

LaToza, T., A. Di Lecce, F. Ricci, B. Towne, and A. van der Hoek. 2015. Ask the Crowd: Scaffolding Coordination and Knowledge Sharing in Microtask Programming. *Proceedings VL/HCC*. Pp. 23–27.

Lentejas, R. 2013. 5 Benefits Businesses Can Get from Consumer Product Reviews. Postsckrippt, February 28. http://www.postsckrippt.ca/5-benefits-businesses-consumer-product-reviews.

Leontiev, A. 1978. *Activity, Consciousness, and Personality*. Englewood Cliffs, NJ: Prentice-Hall. (Originally published in Russian, 1975.)

Letts, G. 2013. The Top 10 Benefits of Customer Feedback. Customersure, July 12. http://www.customersure.com/blog/top-10-benefits-of-customer-feedback.

Levinson, S., and N. Enfield. 2006. *Roots of Human Sociality: Culture, Cognition, and Interaction*. London: Berg Publishers.

Licklider, J. C. R. 1960. Man-Computer Symbiosis. *IRE Transactions on Human Factors in Electronics* HFE-1:4–11.

Maestri, L. and Wakkary, R. 2011. Understanding Repair as a Creative Process of Everyday Design. *Creativity and Cognition*: 81–90.

MacKenzie, D. n.d. Is Economic Performative?: Option Theory and the Construction of Derivatives Markets. http://www.lse.ac.uk/accounting/CARR/pdf/MacKenzie.pdf.

Major, N., and B. Nardi. (in preparation). The Quest for Authenticity. Amateur Content Creators and YouTube.

Manjoo, Farhad. 2015. Uber's Business Model Could Change Your Work, *New York Times*, January 28. http://www.nytimes.com/2015/01/29/technology/personaltech/uber-a-rising-business-model.html

Marazzi, C. 2007. *Capital and Affects: The Politics of the Language Economy*. Los Angeles: Semiotext(e).

Marcuse, H. 1964. *One-Dimensional Man*. Boston: Beacon Press.

Marshall, A. 1890. *The Principles of Economics*. Macmillan and Co.

Martin, D., B. Hanrahan, J. O'Neill, and N. Gupta. 2014. Being a Turker. *Proceedings CSCW* 14:224–235.

Marx, K. 1844. *Economic and Philosophical Manuscripts*.

Marx, K. 1939/1993. *Grundrisse: Foundations of the Critique of Political Economy*. New York: Penguin Classics.

Marx, K. 1990. Vol. I. *Capital: A Critique of Political Economy*. New York: Penguin Classics.

Maxwell, R., ed. 2016. *The Routledge Companion to Labor and Media*. New York: Routledge.

McChesney, R. 2013. *Digital Disconnect: How Capitalism is Turning the Internet Against Democracy*. New York: The New Press.

McCloskey, D. 2009. The Anti-Materialist Project of "The Bourgeois Era." https://mpra.ub.uni-muenchen.de/17411/1/MPRA_paper_17411.pdf.

McCullagh, D. 2010. Why No One Cares about Privacy Anymore? CNET.com, March 12. https://www.cnet.com/news/why-no-one-cares-about-privacy-anymore.

Michelucci, P. 2013. Synthesis and Taxonomy of Human Computation. In *Handbook of Human Computation*, ed. P. Michelucci, 83–86. New York: Springer.

Miller, P., and N. Rose. 2008. *Governing the Present*. London: Polity Press.

Mirowski, P. 2002. *Machine Dreams: Economics Becomes a Cyborg Science*. New York: Cambridge University Press.

Morozov, E. 2011. *The Net Delusion: The Dark Side of Internet Freedom*. New York: Public Affairs.

Moshenska, G. 2007. Unearthing an Air-Raid Shelter at Edgeware Junior School. *London Archaeologist* 11 (9): 237–240.

Murthy, D. 2012. Towards a Sociological Understanding of Social Media: Theorizing Twitter. *Sociology* 46 (6): 1059–1073.

Mutlu, B., and J. Forlizzi. 2008. *Robots in Organizations: The Role of Workflow*. Social and Environmental Factors in Human-Robot Interaction. *Proceedings of the ACM/IEEE International Conference on Human-Robot Interaction (HRI)*.

Nardi, B. 1993. *A Small Matter of Programming: Perspectives on End User Computing*. Cambridge, MA: MIT Press.

Nardi, B. 2010. *My Life as a Night Elf Priest: An Anthropological Account of World of Warcraft*. Ann Arbor: University of Michigan Press.

Nardi, B. 2013. The Role of Human Computation in Sustainability, or, Social Progress Is Made of Fossil Fuels. In *Handbook of Human Computation*. New York: Springer.

Nardi, B. 2015. Inequality and Limits. Special issue "Computing within LIMITS." *First Monday*, August.

Nardi, B., and J. Miller. 1991. Twinkling Lights and Nested Loops: Distributed Problem Solving and Spreadsheet Development. *International Journal of Man-Machine Studies* 34:161–184.

Nardi, B., and V. O'Day. 2000. *Information Ecologies: Using Technology with Heart.* Cambridge, MA: MIT Press.

Nardi, B., D. Schiano, and M. Gumbrecht. 2004. Blogging as Social Activity, or, Would You Let 900 Million People Read Your Diary? *Proceedings Conference on Computer-Supported Cooperative Work.* New York: ACM Press.

Nardi, B., S. Whittaker, and H. Schwarz. 2002. NetWORKers and Their Activity in Intensional Networks. *Computer Supported Cooperative Work* 11:205–242.

Nathan, L. 2008. Ecovillages, Values, and Information Technology: Balancing Sustainability with Daily Life in 21st Century America. *Proceedings of the ACM Conference on Human Factors in Computer Systems, Extended Abstracts.* Pp. 3723–3728.

Neff, G. 2012. *Venture Labor: Work and the Burden of Risk in Innovative Industries.* Cambridge, MA: MIT Press.

North, D. 1990. *Institutions, Institutional Change and Economic Performance.* Cambridge: Cambridge University Press.

O'Connor, S. 2013a. Amazon's Human Robots. *Daily Mail*, February 28. http://www.dailymail.co.uk/news/article-2286227/Amazons-human-robots-Is-future-British-workplace.html.

O'Connor, S. 2013b. Amazon Unpacked. *Financial Times*, February 8. https://www.ft.com/content/ed6a985c-70bd-11e2-85d0-00144feab49a.

Ohm, P., and Peppet, S. 2016. What If Everything Reveals Everything. In *Big Data Is Not a Monolith*, eds. C. Sugimoto, H. Ekbia, and M. Mattioli. Cambridge, MA: MIT Press.

O'Malley, Pat. 1992. Risk, Power and Crime Prevention. *Economy and Society* 21:252–275.

O'Reilly, T. 2005. What Is Web 2.0: Design Patterns and Business Models for the Next Generation of Software. O'Reilly.com http://oreillynet.com/1pt/a/6228.

Oxfam 2016. 62 People Own the Same as Half the World, Reveals Oxfam Davos Report. https://www.oxfam.org/en/pressroom/pressreleases/2016-01-18/62-people-own-same-half-world-reveals-oxfam-davos-report.

Packard, V. 1960. *The Waste Makers.* New York: David Mackay Company.

Paolacci, G., and J. Chandler. 2014. Inside the Turk: Understanding Mechanical Turk as a Participant Pool. *Current Directions in Psychological Science* 23 (3): 184–188.

Pargman, D., and B. Raghavan. 2015. Introduction to LIMITS '15: First Workshop on Computing within Limits. *First Monday*, August.

Patterson, D. 2015. Haitian Resiliency: A Case Study in Intermittent Infrastructure. *First Monday*, August.

Paul, C. 2011. Optimizing Play: How Theorycraft Changes Gameplay and Design. *Games and Culture* 11 (2).

Penny, S. 2016. The Elephants in the (Server) Room: Sustainability and Surveillance in the Era of Big Data. In *Ubiquitous Computing, Complexity and Culture*, eds. U. Ekman, et al. New York: Routledge.

Perry, S., and N. Beale. 2015. The Social Web and Archaeology's Restructuring: Impact, Exploitation, Disciplinary Change. *Open Archaeology* 1:153–165.

Petcou, C., and D. Petrescu. 2014. *R-urban: Strategies and Tactics for Participative Utopias and Resilient Practices. Green Utopianism: Perspectives, Politics and Micro-Practices*, eds. K. Bradley and J. Hedrén. Routledge.

Peters, T. 1992. *Liberation Management*. New York: Alfred A. Knopf.

Peterson, V. 2005. How (the Meaning of) Gender Matters in Political Economy. *New Political Economy* 10 (4): 499–521.

Pierce, J. 2012. Undesigning Technology: Considering the Negation of Design by Design. *Proceedings of the ACM Conference on Human Factors in Computer Systems*. Pp. 957–966.

Piketty, T. 2014. *Capital in the Twenty-First Century*. Cambridge, MA: Belknap Press.

Pinch, T. 2010. Presidential Address, Annual Meeting of the Society for the Social Studies of Science. Tokyo, Japan.

Pinker, S. 2007. *The Language Instinct*. New York: Harper Perennial Modern Classics.

Podolny, S. 2015. If an Algorithm Wrote This, How Would You Even Know? *New York Times*, March 7.

Polanyi, M. 1949. *The Great Transformation: The Political and Economic Origins of Our Time*. New York: Beacon Press.

Postigo, H. 2003. From Pong to Planet Quake: Post Industrial Transitions from Leisure to Work. *Information Communication and Society* 6 (4): 593–607.

Postigo, H. 2010. Modding to the Big Leagues: Exploring the Space between Modders and the Game Industry. *First Monday*, May.

Proffitt, J., H. R. Ekbia, and S. McDowell. 2015. Introduction to the Special Forum on Monetization of User-Generated Content—Marx Revisited. *Information Society* 31 (1): 1–4.

Qaurooni, D., A. Ghazinejad, I. Kouper, and H. R. Ekbia. 2016. Citizens for Science and Science for Citizens: The View from Participatory Design. *Proceedings of the ACM Conference on Human Factors in Computer Systems*.

Raghavan, R., and S. Hasan. 2012. Macroscopically Sustainable Networking: An Internet Quine. ICSI Technical Report, September. Berkeley: ICSI.

Raghavan, B., B. Nardi, S. Lovell, J. Norton, B. Tomlinson, and D. Patterson, 2016. Computational Agroecology: Sustainable Food Ecosystem Design. *Proceedings CHI'2016*.

Raval, N., and P. Dourish. 2016. Standing Out from the Crowd: Emotional Labor, Body Labor, and Temporal Labor in Ridesharing. *Proceedings CSCW*:97–107.

Ratto, M. 2011. Critical Making: Conceptual and Material Studies in Technology and Social Life. *Information Society* 27 (4): 252–260.

Reich, R. 2015. The Sharing Economy Is Hurtling Us Backwards. *Salon*, February 4. http://www.salon.com/2015/02/04/robert_reich_the_sharing_economy_is_hurtling_us_backwards_partner.

Remy, C., and E. Huang. 2014. Addressing the Obsolescence of End-User Devices: Approaches from the Field of Sustainable HCI. In *ICT Innovations for Sustainability*. New York: Springer.

Rheingold, H. 2000. *The Virtual Community: Homesteading on the Electronic Frontier*. Cambridge, MA: MIT Press.

Rifkin, J. 1995. *The End of Work*. New York: Putnam.

Rifkin, J. 2000. *The Age of Access: The New Culture of Hypercapitalism*. New York: Tarcher/Putnam.

Rifkin, J. 2014. *The Zero Marginal Cost Society*. New York: Palgrave Macmillan.

Ritzer, G., and N. Jurgenson. 2010. Production, Consumption, Prosumption: The Nature of Capitalism in the Age of the Digital "Prosumer." *Journal of Consumer Culture* 10 (1): 13–36.

Robertson, M. 2012. Measurement and Alienation: Making a World of Ecosystem Services. *Transactions of the Institute of British Geographers* 37 (3): 386–401.

Robinson, B. 2015. With a Different Marx: Value and the Contradictions of Web 2.0 Capitalism. *Information Society* 31 (1): 44–51.

Robinson, M. 1980. *Housekeeping*. New York: Farrar, Straus, and Giroux.

Rooney, B. 2014. The FedEx Driver Who Sued and Won. CNN Money, November 21. http://money.cnn.com/2014/11/20/news/companies/fedex-driver-lawsuit.

Rose, N., and J. Abi-Rached. 2013. *Neuro: The New Brain Sciences and the Management of the Mind*. Princeton, NJ: Princeton University Press.

Saez, E., and G. Zucman. 2014. Wealth Inequality in the United States since 1913. http://gabriel-zucman.eu/files/SaezZucman2014Slides.pdf.

Schiano, D. 1999. Lessons from Lambda Moo. *Presence* 8 (2): 127–139.

Schiller, D. 1999. *Digital Capitalism*. Cambridge, MA: MIT Press.

Schiller, D. 2014. *Digital Depresssion: Information Technology and Economic Crisis*. Urbana: University of Illinois Press.

Schmidt, E., and J. Cohen. 2014. *The New Digital Age: Reshaping the Future of People, Nations, and Business*. New York: Vintage Books.

Schmidt, F. 2013a. For a Few Dollars More: Class Action Against Crowdsourcing. #BWPWAP 2.1. http://www.aprja.net/?p=836.

Schmidt, F. 2013b. The Good, the Bad and the Ugly. IEEE Workshop Cloud and Green Computing. http://florianschmidt.co/the-good-the-bad-and-the-ugly.

Scholz, T. 2013. *Digital Labour: The Internet as Playground and Factory*. New York: Routledge.

Schumacher, E. F. 1973. *Small Is Beautiful: A Study of Economics As If People Mattered*. London: Blond and Briggs.

Schumpeter, J. 1942. *Capitalism, Socialism, and Democracy*. London: Routledge.

Senior, T. 2011. Riot Games Hopes Tribunal System Will Clean Up League of Legends Community. http://www.pcgamer.com/2011/01/14/riot-games-hopes-tribunal-system-will-clean-up-league-of-legends-community.

Sennett, R. 2007. *The Culture of the New Capitalism*. New Haven, CT: Yale University Press.

Sennett, R. 2012. *Together: The Rituals, Pleasures, and Politics of Cooperation*. New Haven, CT: Yale University Press.

Shapiro, C., and H. L. Varian. 1998. *Information Rules: A Strategic Guide to the Network Economy*. Boston, MA: Harvard Business School Press.

Shibata, J. 2012. Therapeutic Seal Robot as Biofeedback Medical Device: Qualitative and Quantitative Evaluations of Robot Therapy in Dementia Care. *Proceedings of the IEEE* 100 (8): 2527–2538.

Shilton, K. 2016. When They Are Your Big Data: Participatory Data Practices as a Lens on Big Data. In *Big Data Is Not a Monolith*, ed. C. Sugimoto, H. Ekbia, and M. Mattioli, 21–30. Cambridge, MA: MIT Press.

Shirky, C. 2010. *Cognitive Surplus: Creativity and Generosity in a Connected Age*. London: Penguin Press.

Silberman, M. S. 2015. Human-Centered Computing and the Future of Work: Lessons from Mechanical Turk and Turkopticon, 2008–2015. PhD thesis, University of California, Irvine.

Simon, H. A. 1955. A Behavioral Model of Rational Choice. *Quarterly Journal of Economics* 69:99–118.

Simon, H. A. 1960. The Corporation: Will It Be Managed by Machines? In *Management and the Corporations 1985*, ed. M. L. Anshen and G. L. Bach, 17–55. New York: McGraw-Hill.

Simon, H. A. 1973. The Organization of Complex Systems. In *Hierarchy Theory*, ed. H. Patee, 3–27. New York: Braziller.

Simon, H. A. 1977. *The New Science of Management Decision*. Englewood, NJ: Prentice-Hall.

Simon, H. A. 1978. *Rational Decision-Making in Business Organizations*. Nobel Memorial Lecture.

Simon, H. A. 1982. *The Sciences of the Artificial*. Cambridge, MA: MIT Press.

Simon, H. A. 1987. Two Heads Are Better than One: The Collaboration between AI and OR. *Interfaces* 17 (4): 8–15.

Simpson, F., and H. Williams. 2008. Evaluating Community Archaeology in the UK. *Public Archaeology* 7 (2): 69–90.

Smith, A. 1776. *An Inquiry into the Nature and Causes of Wealth of Nations*.

Smythe, D. W. 1981. *Dependency Road: Communications, Capitalism, Consciousness and Canada*. Norwood, NJ: Ablex Publishing.

Snider, L. 2014. Interrogating the Algorithm: Debt, Derivatives and the Social Reconstruction of Stock Market Trading. *Critical Sociology* 40 (5): 747–761.

Sood, S., J. Antin, and E. Churchill. 2012. Profanity Use in Online Communities. *Proceedings of the ACM Conference on Human Factors in Computer Systems.* Pp. 1481–1490.

Sotamaa, O. 2009. The Player's Game: Towards Understanding Player Production among Computer Game Cultures. Ph.D. thesis, Tampere, Finland: University of Tampere.

Sproull, L., and S. Kiesler. 1986. Reducing Social Context Cues: Electronic Mail in Organizational Communication. *Management Science* 32 (11): 1492–1512.

Sprouse, J. 2011. A Validation of Amazon Mechanical Turk for the Collection of Acceptability Judgments in Linguistic Theory. *Behavior Research Methods* 43 (1): 155–167.

Standing, G. 2011. *The Precariat.* London: Bloomsbury Academic.

Star, S. L. 1990. Power, Technology and the Phenomenology of Conventions: On Being Allergic to Onions. *Sociological Review* 38 (1): 26–56.

Stiglitz, J. 2006. *Making Globalization Work.* New York: W. W. Norton.

Stiglitz, J. 2014. Is Inequality Inevitable? *New York Times,* June 29.

Stone, K. V. W. 2006. Flexibilization, Globalization, and Privatization: Three Challenges to Labor Rights in Our Time. *Osgoode Hall Law Journal* 44 (1): 77–104.

Suarez-Villa, L. 2012. *Globalization and Technocapitalism: The Political Economy of Corporate Power and Technological Domination.* Farnham: Ashgate.

Suarez-Villa, L. 2015. *Corporate Power, Oligopolies, and the Crisis of the State.* Albany: SUNY Press.

Suler, J. 2004. The Online Disinhibition Effect. The Impact of the Internet, Multimedia and Virtual Reality on Behavior and Society. *Cyberpsychology & Behavior* 7 (3): 321–326.

Sung, J.-Y., L. Guo, R. Grinter, and H. Christensen. 2007. "My Roomba Is Rambo": Intimate Home Appliances. *Proceedings of the International Conference on Ubiquitous Computing.*

Swallow, S. 2011. How Recruiters Use Social Networks to Screen Candidates. *Mashable,* October 23. http://mashable.com/2011/10/23/how-recruiters-use-social-networks-to-screen-candidates-infographic.

Tainter, J. 1990. *The Collapse of Complex Societies.* Cambridge: Cambridge University Press.

Takeuchi, Y. 2016. Printable Hydroponic Gardens: Initial Explorations and Considerations. *Proceedings of the ACM Conference on Human Factors in Computer Systems,* 449–458. New York: ACM.

Tanaka, F., A. Cicourel, and J. Movellan. 2007. Socialization between Toddlers and Robots at an Early Childhood Education Center. *Proceedings of the National Academy of Sciences of the United States of America (PNAS)* 104 (46): 17954–17958.

Tanenbaum, J., A. Desjardins, and K. Tanenbaum. 2013. Steampunking Interaction Design. *Interactions* (May–June): 28–30.

Targett, S., V. Verlysdonk, H. Hamilton, and D. Hepting. 2012. A Study of Interface Modifications in World of Warcraft. *Game Studies* 12 (2).

Taylor, A. 2014. *The People's Platform: Taking Back Power and Culture in the Digital Age*. New York: Metropolitan Books.

Taylor, T. L. 2006a. Does WoW Change Everything?: How a PVP Server, Multinational Player Base, and Surveillance Mod Scene Caused Me Pause. *Games and Culture* 1 (4): 318–337.

Taylor, T. L. 2006b. *Play between Worlds: Exploring Online Game Culture*. Cambridge, MA: MIT Press.

Terranova, T. 2000. Free Labor: Producing Culture for the Digital Economy. *Social Text* 18 (2): 33–58.

Terranova, T. 2003. Free Labor: Producing Culture for the Digital Economy. *Electronic Book Review*, June 20. http://www.electronicbookreview.com/thread/technocapitalism/voluntary.

Thornton, R. 1847. *The Expounder of Primitive Christianity*. Ann Arbor, MI.

Thrift, N. 2005. *Knowing Capitalism*. London: Sage.

Tsui, K., M. Desai, H. Yanco, and C. Uhlik. 2011. Exploring Use Cases for Telepresence Robots. *Proceedings of the 6th ACM/IEEE International Conference on Human-Robot Interaction*.

Turkle, S. 1997. *Life on the Screen: Identity in the Age of the Internet*. New York: Simon and Schuster.

Turkle, S. 2011. *Alone Together: Why We Expect More from Technology and Less from Each Other*. New York: Basic Books.

Ueno, N., R. Sawyer, and Y. Moro. 2016. Artifacts, Agency and Socio-technical Arrangement. *Journal of Mind, Culture and Activity* (July).

US Bureau of Census. 1976. *Historical Statistics in the United States*. Series F1–F5.

US Bureau of Census. 2010. *Briefs*.

Vandermeer, J. 2011. *The Ecology of Agroecosystems*. New York: Jones & Bartlett.

van Dijk, J. 2009. Users like you? Theorizing Agency in User-Generated Content. *Media Culture & Society* 31:41–58.

van Dijk, J. 2013. *The Culture of Connectivity. A Critical History of Social Media*. Oxford: Oxford University Press.

van Dijk, J., and D. Nieborg. 2009. Wikinomics and Its Discontents: A Critical Analysis of Web 2.0 Business Manifestos. *New Media & Society* 11 (5): 855–874.

Vercellone, C. 2007. From Formal Subsumption to General Intellect: Elements for a Marxist Reading to the Thesis of Cognitive Capitalism. *Historical Materialism* 15: 13–36.

Vermuelen, S., B. Campbell, and J. Ingram. 2012. Climate Change and Food Systems. *Annual Review of Environment and Resources* 37 (1): 195–222.

Vertesi, J. 2012. Seeing Like a Rover: Visualization, Embodiment, and Interaction on the Mars Exploration Rover Mission. *Social Studies of Science* 42:393–414.

Virilio, P. 1986. *Speed and Politics: An Essay on Dromology.* New York: Semiotext(e).

Vygotsky, L. 1986. *Thought and Language.* Cambridge, MA: MIT Press.

Wada, K., Y. Ikeda, K. Inoue, and R. Uehara. 2010. Development and Preliminary Evaluation of a Caregiver's Manual for Robot Therapy using the Therapeutic Seal Robot Paro. Paper presented at the RO-MAN Conference.

Wada K., and T. Shibata. 2007. Living with Seal Robots–Its Socio-Psychological and Physiological Influences on the Elderly in a Care House. *IEEE Trans Robot* 23:972–980.

Wada, K., T. Shibata, T. Saito, K. Sakamoto, and Kazuo Tanie. 2005. Psychological and Social Effects of One-Year Robot Assisted Activity on Elderly People at a Health Service Facility for the Aged. *Proceedings of the 2005 IEEE International Conference on Robotics and Automation.* Pp. 2796–2801.

Wagner, P. 1994. *A Sociology of Modernity: Liberty and Discipline.* London: Routledge.

Wagner, C. 2004. Wiki: A Technology for Conversational Knowledge Management and Group Collaboration. *Communications of the Association for Information Systems* 13 (Article 19).

Wallerstein, I. 1983. *Historical Capitalism.* New York: Verso.

Want, R. 2001. Ten Lessons Learned about Ubiquitous Computing. https://www.vs.inf.ethz.ch/events/dag2001/slides/roy-lessons.pdf

Want, R., A. Hopper, V. Falcão, and J. Gibbons. 1992. The Active Badge Location System. *ACM Transactions on Information Systems (TOIS)* 10 (1): 91–102.

Waterman, A. n.d. Neoclassical and Classical Growth Theory Compared. http://amcwaterman.com/working-papers.

Webb, C. 2004. Google's Eyes in Your Inbox. *Washington Post.* http://www.washingtonpost.com/wp-dyn/articles/A44454-2004Apr2.html.

Weber, L., and M. Korn. 2014. Where Did All the Entry-Level Jobs Go? Many Firms Expect New Graduates to Arrive Job-Ready from Day One. *Wall Street Journal*, August 6.

Weber, M. 1904. *The Protestant Ethic and the Spirit of Capitalism.* London: Unwin Hyman.

Webster, F. 1995. *Theories of the Information Society.* New York: Routledge.

Weeks, K. 2011. *The Problem with Work: Feminism, Marxism, Antiwork Politics, and Postwork Imaginaries.* Durham, NC: Duke University Press.

Williamson, O. 1985. *The Economic Institutions of Capitalism.* New York: The Free Press.

Winner, L. 1986. *The Whale and the Reactor: A Search for Limits in an Age of High Technology.* Chicago: University of Chicago Press.

Woelfer, J., and D. Hendry. 2011. Homeless Young People and Technology. *Interaction* (November–December):70–73.

Womack, J., D. Jones, and D. Roos. 2007. *The Machine That Changed the World: The Story of Lean Production: Toyota's Secret Weapon in the Global Car Wars That Is Revolutionizing World Industry*. New York: Free Press.

World Economic Forum. 2011. Unlocking the Value of Personal Data: From Collection to Usage. http://www.weforum.org/reports/unlocking-value-personal-data-collection-usage.

Wozniak, S. 2011. Hungry for AI Era. http://www.sbs.com.au/news/article/1694857/Wozniak-hungry-for-AI-era.

Wright, E. O. 1997. *Class Counts: Comparative Studies in Class Analysis*. New York: Cambridge University Press.

Wright, E. O. 2005. Foundations of a Neo-Marxist Class Analysis. In *Approaches to Class Analysis*, ed. E. O. Wright, 1–26. New York: Cambridge University Press.

Wright, E. O. 2009. *Envisioning Real Utopias*. New York: Verso.

Zhou, Y. 2005. Living on the Cyber Border: "Minjian" Political Writers in Chinese Cyberspace. *Current Anthropology* 46 (5): 779–803.

Zittrain, J. 2008. *The Future of the Internet*. London: Allen Lane.

Zuboff, S. 1988. *In the Age of the Smart Machine*. New York: Basic Books.

Index

ABC Television Network, 67
Active Badges, 76–77
Activity theory, 38
Accumulation of capital, 1, 11, 17, 32, 44–56, 71, 73, 185, 187–188, 209
  by dispossession, 45, 56, 187, 192, 194, 208
  of wealth, 18, 39, 45, 47, 52–53, 56, 71, 82, 170–171, 188–190, 207
Acxiom, 94, 97
Addiction, 11, 167
Advertising, 35, 38, 52, 60, 72, 94–95
Affective capitalism, 50, 95, 120
  value-affect, 50
Affordable Care Act, 85, 179. *See also* Health insurance
Airbnb, 184
Alač, Morana, 131
Alaimo, Cristina, 98
Alcott, Louisa May, 83
Alienation, 32, 43, 83, 87–88, 102, 116, 139, 157, 168, 176
Altieri, Miguel, 197, 221
Amazon, 12, 31, 38, 60, 82, 107, 113, 121–122, 138, 142, 144, 155, 166, 174, 182–184, 203, 218
  Mechanical Turk, 12, 38, 60, 107, 113–116, 121–122, 138, 142, 144, 166, 174, 184, 203, 218
Analytical Engine, 23
André, Paul, 35
Andrejevic, Mark, 61, 95, 97
Antin, Judd, 84

Anxieties, 102, 115–116, 154, 157, 169, 176, 179, 212, 223. *See also* Alienation
Apple Computer, 27–30
Arendt, Hannah, 41
ARPAnet, 77
Arrow, Kenneth, 47
Artificial Intelligence (AI), 2, 13, 26, 29, 36, 77, 107, 129, 146, 181, 183, 222
Arvidsson, Adam, 12, 50, 93
Ashby, W. Ross, 26
Aspray, William, 24, 28–29
Assange, Julian, 221
Association of Internet Researchers (AoIR), 31
AT&T, 29
Auger, James, 130
Augmentation, 24, 26–32, 39, 57, 78, 115
Automated Teller Machine (ATM), 18, 60, 108–109, 140–141, 164
Automation, 19, 24–26, 32, 38, 39, 51–52, 56, 57–59, 69, 98, 108–116, 170, 180, 181, 188, 205, 207, 219, 220, 222
Autonomy, 45, 50–51, 62, 75, 78, 130–146, 164, 179, 185, 209

Babbage, Charles, 9, 23
Bailey, Diane, 19, 126, 140–142, 144
Bank Innovation, 109
Barabási, Albert-László, 10

Basic guaranteed income, 188, 198, 203, 207, 209
Baumer, Eric, 198
Baxter, 180–182. *See also* Industrial robots
Bear-Stearns, 84
Beck, Ulrich, 74, 81, 84
Becker, Gary, 154, 222
Beer, David, 101
Beniger, James, 25
Benkler, Yochai, 48
Berardi, Franco, 50–51
Bezos, Jeff, 156
Black, Bob, 203
Blevis, Eli, 14, 198
Blizzard Entertainment, 62, 120–122, 125
Bollier, David, 48
Boltanski, Luc, 10–11, 23, 30–31, 64, 78, 81, 86, 99
Böhm-Bawerk, 45
Booth, Douglas, 188
Bradley, Karin, 199, 221
Braudel, Fernand, 209
Breivik, Anders, 167
Bricklin, Dan, 27
Broekens, Joost, 130
Brooks, Rodney, 181–183
Brown, Wendy, 190
Bruckman, Amy, 154
Brynjarsdóttir, Hrönn, 198
Brynjolfsson, Erik, 169–170, 180–183, 207, 213–214, 220, 222
Bubble-stars, 171–174, 176
Buchanan, James, 46
Buchsbaum, Jonathan, 124
Bullard, Julia, 123
Burawoy, Michael, 33–34, 55, 57, 147
Burdon, Mark, 97
Bureaucracy, 24, 65, 72, 148, 154
Burkett, Paul, 188
Bush, Vannevar, 7

Casualization, 73, 75, 203
Caffentzis, George, 11, 14, 24, 57–62, 70–71, 81, 176, 200, 204
Callois, Roger, 121
Campbell-Kelly, Martin, 24, 28–29, 31
Cappella, Joseph, 172
Caring capacity, 32, 38, 139–140, 146
Castells, Manuel, 10, 101
CBS Television Network, 67, 68
Cerrato-Pargman, Tessy, 198
Cervantes, Ruy, 134–135
Chandler, Jesse, 113
Chang, Wan-Ling, 37, 130–135
Chaplin, Charlie, 217
Chiapello, Ève, 10–11, 23, 30–31, 64, 78, 81, 86, 99
Chilton, Lydia, 60
Chootanom, Trina, 123
Churchill, Elizabeth, 84
Citizen science, 10, 12, 32, 38, 150–152, 166, 174, 176, 203, 218
Class formation, 54, 55
Class structures, 54, 55, 201
Clean Air Act, 73. *See also* Clean Water Act *and* Endangered Species Act
Clean Water Act, 73. *See also* Clean Air Act *and* Endangered Species Act
Climate change, 197, 206
Coase, Ronald, 47
Coercion, 4, 8, 52, 64–66, 109, 111, 148–156, 220. *See also* Invisible control
 coercive mechanisms, 4, 52, 64, 66, 109, 149, 152
Cohen, Roger, 221
Cohen, Stanley, 167
Colleoni, Elanor, 12, 19, 26, 50, 93, 140–142
Collins, Harry, 11
Communicative labor, 18, 89, 96
Comor, Edward, 12, 61, 97
Conaty, Patrick, 48
Contingent labor, 30
Control mechanisms, 14–15, 34, 52, 55, 64–66, 76, 79, 112, 126, 147–148, 161, 188, 213, 215
 consent-producing practices, 14, 17, 65–66, 79, 148 (*see also* Coercion)

Index   259

Convenience, 42, 56, 99, 108–116, 161, 164–165, 213
Cognitive labor, 18, 32–33, 37, 89, 107–116, 141
Correspondent banking, 140–146, 174, 176, 218
Council of Economic Advisors, 184
Country Joe and the Fish, 69–70
Creative labor, 19, 37, 117–127
Croll, Alistair, 180
CrowdFlower, 12, 113
Crowdsourcing, 12, 33–35, 47–48, 126–127, 143, 152, 175
Cultural influences, 9, 38, 61, 66–72, 87, 117, 168–169, 190, 208
  counterculture, 68–72
Customer reviews, 149, 152

Daft, Richard, 149
Daly, Herman, 188
Dark Mountain Collective, 208, 221
Darrach, Brad, 222
Darwin, Charles, 17
Dean, Howard, 152
De Brunhoff, Suzanne, 73, 75
Debt, 76, 86, 123, 141, 148, 200
Deleuze, Gilles, 15, 83, 176
Deregulation, 73–74, 180, 222
Dertouzos, Michael, 26
Desjardins, Audrey, 198
De Stefano, Valerio, 203
Dewey, John, 84, 88, 166
Dialectic, 7, 9, 39, 43–45, 56, 64
Dick, Philip K., 58–59, 130
Dickens, Charles, 176, 217
Digital labor, 13, 48–51, 170–173
Dillahunt, Tawanna, 198
Diniz, Eduardo, 19, 126, 140–142
DiSalvo, Carl, 130
Disneyland, 122
Displacement, 18
Dispossession, 44–45, 56, 187, 192, 208
Dissent, 16, 67–68, 200, 220
Division of labor, 9, 32–33, 37, 42–43, 131–132, 147, 153, 186, 188, 196, 203, 206, 208

Division of power, 126, 152–157, 161, 170, 188, 198, 203, 206, 208
Dunne, Anthony, 13
Dye-Witheford, Nick, 221–222

Ecology, 46, 188, 190, 197–198, 208
E-commerce, 31, 93, 149
Economic inequality, 153, 157, 170–171, 185, 187, 190, 198, 206, 207, 214–216, 223
Eisenhower, Dwight D., 70
Electronic banking, 31, 108–109, 140–146, 218
Elgin, Duane, 73
Eliza program, 136
El-Ouazzane, Remi, 183
Emirbayer, Mustafa, 37
Employment, 9, 18, 24, 30–32, 46, 55, 75, 84–89, 107–116, 171, 178, 180–186, 203, 212, 220
  contract, 30, 55, 75, 113, 122–123, 180–183, 185
  low-paid or low-waged, 24, 55, 62, 75, 84–85, 113, 138, 144, 178, 180–183, 220
  precarious, 18, 30–32, 55, 75–76, 81, 84–86, 94, 122–127, 171–173, 178, 180–183, 212
  secure, 30, 46, 55, 57, 62, 76, 86, 178, 180, 184, 203, 209, 212, 220
  self-employment, 30–31, 75, 85, 122–123, 171–173, 178, 184
  unemployment and underemployment, 84, 86, 89, 124, 139, 178, 180–184, 220
Emotional labor, 19, 32, 89, 129–146
Endangered Species Act, 73. *See also* Clean Air Act *and* Clean Water Act
Enfield, NJ, 93
Engelbart, Douglas, 26
Engels, Friedrich, 2
Environment, 46, 188, 196, 200, 206, 209, 221
Environmental Protection Act (EPA), 73
Enron, 84
Epson, 97

Ethnicity, 9, 68
Etzioni, Amitai, 204, 206
Exploitation, 11, 16, 44–45, 48–49, 55, 85–87, 94–95, 187–188, 194, 196, 209, 217–219

Facebook, 49, 55, 82, 93–94, 96, 98–101, 104, 111, 155, 173, 174, 212, 218, 220
Federici, Silvia, 46, 61
FedEx, 182
Fehr, Ernst, 154
Fergusson, Robert, 83, 102, 196
Ferlinghetti, Lawrence, 68
Feudalism, 14, 34, 57, 65, 147
Fisher, David, 67
Fisher, Marshall Jon, 67
Flexibilization, 75, 124, 182, 184–186, 203
Flickr, 100, 153
Food production, 196–201, 205, 209
Forte, Andrea, 154
Foucault, Michel, 3, 8, 11, 16, 41, 63–64, 74, 79, 100, 164–169, 177, 211, 215
Forbes, 102
Forlizzi, Jodi, 130
Foster, John Bellamy, 46
Francis, Anne, 87
Franklin, Benjamin, 23
Frankston, Bob, 27
Freire, Paulo, 3
Friedman, Thomas, 46, 64
Fuchs, Christian, 12, 13, 49, 50, 93

Galbraith, John Kenneth, 46
Galloway, James Cole, 49
Gassé, John-Louis, 27
Gates, Bill, 129, 156
Gender, 9, 46, 61, 84
General Motors, 51
Georgescu-Roegen, Nicholas, 187
Giddens, Anthony, 81
Ginsberg, Allen, 68
Globalization, 124, 171, 180, 182, 190, 201, 207
Goldberg, Ivan, 166

Goldcorp, Inc., 155–156
Gollac, Michel, 30
Google, 55, 78, 82, 97, 102, 106, 155, 156, 174, 218, 219
Gordon, Robert, 75, 207
Gorz, André, 81, 188–190, 202, 206–209
Goulding, Christina, 117
Gramsci, Antonio, 65
Graphic design, 121–122, 126–127, 221
Great Depression, 72
Great Recession, 183
Greenwich Village, 69
Grudin, Jonathan, 29, 222–223
Gui, Xinning, 199

Haight-Ashbury, 69
Håkansson, Maria, 198
Handy, 182
Harder, Delmar, 25
Hardt, Michel, 50
Harvey, David, 6, 11, 23, 44, 55, 65, 81
Hasan, Shaddi, 202
Haushofer, Johannes, 154
Hawthorne, Nathaniel, 83
Hayden-Smith, Rose, 201
Healthcare, 46, 85, 177–179, 181, 190, 212
Health insurance, 85, 177–179, 218. *See also* Affordable Care Act
Health IQ, Inc., 166, 174–176, 180
Heerink, Marcel, 130
Heidegger, Martin, 37, 41
Hellman, Matilda, 167
Hendry, David, 18
Heston, Charlton, 173
Hewlett Packard, 27, 29
Hidden information, 47, 59, 61–62, 175–176, 219
Hidden labor, 35, 61–62, 95–96, 114, 126, 129–146, 157, 174, 180–183, 219
Hidden value, 34, 59, 61–62, 95–96, 99, 112, 114, 124, 139–140, 175–176, 219

Index 261

Hierarchy, 69, 147–148, 152, 156
Hollywood, 2, 67, 105, 146, 172–173
Holmes, Brian, 3, 81
Hornborg, Alf, 8, 81, 187, 221
Hospitality industry, 114
Howard, Robert, 69
Huang, Elaine, 14
Huber, Lesa, 130
Huff, Darrell, 156
Hughes Aircraft, 75
Human Computation, 28, 32–34, 114, 126, 152, 157
Human Intelligence Tasks (HITs), 113, 121, 138, 156
Huffman, Mark, 110–111
Huizinga, Johan, 121
Huntemann, Aslinger, 84
Hüttenraunch, Helge, 130
Huws, Ursula, 46

Iacano, Suzie, 222–223
IBM Corporation (International Business Machines), 29, 110
Identity, 4, 76, 82–83, 94, 117–118, 121–127, 168–169, 172, 208
Industrialism, 41, 147–148, 157, 178, 197, 218
Informatics, 12
Innovation, 6, 41, 156
Inoue, Kaoru, 130
Instacart, 184
Instagram, 170–171, 218
Intel, 29
Intellicorp, 77
Intrinsic motivation, 148, 154, 204, 218, 220
Invisible control, 39, 99, 126, 152, 214
Invisible value, 1, 17, 95, 99, 114, 124, 152–157, 175–176. *See also* Hidden value
Irani, Lilly, 60, 180
Irving, Washington, 83
Ito, Yuko, 130

Jackson, Steven, 198
Jacquard, Joseph, 23

Jacquard loom, 24–25
James, Henry, 83
Jameson, Fredric, 65
Jamieson, Kathleen Hall, 172
JC Penney, 75
Jiang, Ling, 60, 166
Jobs, Steve, 27–28
John, N. A., 98
Johnson & Johnson, 75
Johnson, Lyndon B., 72
Jones, Daniel, 148
Joshi, Somya, 198
Jurgenson, Nathan, 12, 35, 111

Kaldor, Nicholas, 45
Kallinikos, Jannis, 98, 219
Kanda, Takayuki, 130
Kaptelinin, Victor, 37
Keep Our Pension Promises Act (KOPPA), 183
Kessler, Sarah, 182
Kiesler, Sara, 83, 84
King, John Leslie, 222–223
Kirman, Ben, 13
Kiva, 182–183. *See also* Amazon; Robotics
Klein, Naomi, 221
Kling, Rob, 14, 222
Klout, 94, 101
Knoblauch, Hubert, 117
Knowles, Bran, 198–199
Krueger, Alan, 184
Kou, Yubo, 35, 112
Körner, Astrid, 154
Kow, Yong Ming, 112, 118–122, 125
Kücklich, Julian, 118, 126
Kumar, Anjali, 40–142
Kunda, Ziva, 30

Labor-power, 61–62
Labor-time, 57, 62, 82, 94, 142, 164, 208, 209
Labor unions, 52, 139, 180
Laney, Doug, 94
Langlois, Richard, 32, 53, 213
Larco, Vanesa, 154
LaToza, Thomas, 207

League of Legends, 107, 111–113, 114, 115
Lentejas, Rodolfo, 149
Leontief, Wassily, 222
Leontiev, A. N., 37
Letts, Guy, 149
Levinson, Steven, 93
Liberalism, 63–64
Licklider, J.C.R., 26
LinkedIn, 93–94, 99–100, 212
Linux, 47, 152, 153, 155
Lovell, Sarah, 196
Lyft, 182

MacKenzie, Donald, 53
Macy's, 97
Madonna, 100
Maestri, Leah, 198
Major, Nathan, 103
Manjoo, Farhad, 184
Manufacturing, 7, 9, 23, 28, 62, 67, 106, 239, 202–206, 220
Marazzi, Christian, 50
Massachusetts Institute of Technology (MIT), 117, 181
Masses, 9–10, 14, 31, 59, 67–68, 81, 99, 105, 117, 144, 172–176, 188, 197, 204, 205, 208
Massification, 67–68, 72, 87, 172, 198, 199
Marcuse, Herbert, 67, 217
Market, 45, 63, 67, 71, 78, 184
Marshall, Alfred, 45
Marx, Karl, 2–3, 11, 16, 24, 32–34, 43–45, 53, 55, 57–61, 97, 114, 124, 198, 200, 209, 211, 215
Marxism, 2–3, 13, 32, 43–45, 54, 58, 94, 188
Maxwell, Richard, 12
McAfee, Andrew, 169–170, 180–183, 207, 213–214, 220, 222
McChesney, Robert, 13, 52
McCloskey, Deirdre, 6
McCullagh, Declan, 100
McDowell, Stephen, 13
McGuire, Timothy, 83
McNealy, Scott, 77

Metaphor, Computer, 76
Microsoft, 29
Michelucci, Pietro, 48, 157
Microvalidations, 165–166, 168–169, 175–176
Microwork, 12, 17–18, 113, 114, 174, 203
Mill, John Stuart, 42
Miller, James, 28
Miller, Peter, 179
Minimum wage, 60, 145, 182–183, 203
Minsky, Marvin, 222
Mirowski, Philip, 46, 53
Mische, Ann, 37
Mitchell, Arnold, 73
Modernism, 2, 63, 82, 100
Modifications to games (mods), 118–127, 171–173, 176, 203
Moro, Yuji, 199
Morozov, Evgeny, 23, 190
Movellan, Javier, 131
Multi-Employer Pension Recovery Act (MPRA), 183
Murthy, Dhiraj, 168
Mutlu, Bilge, 130

Nathan, Lisa, 198
National Aeronautics and Space Administration (NASA), 73, 131, 205
National Oceanic and Atmospheric Administration (NOAA), 73
National Public Radio (NPR), 109
National Rifle Association (NRA), 173
NBC Television Network, 67
Negri, Antonio, 50
Neoconservatism, 85, 179, 185
Neoliberalism, 9, 46, 61, 65, 74, 82, 85, 88, 177, 179, 204, 215–216, 222
Neoclassicism, 45–47, 52–53, 188
Neff, Gina, 84
New Deal, 68
Nieborg, David, 35

Index 263

Nixon, Richard M., 73
North, Douglass, 47, 53

Obama, Barack, 85, 179, 184
Offshoring, 30
Ohm, Paul, 179
Olivetti, S.p.A., 76
O'Day, Vicki, 222
O'Connor, Sarah, 182
O'Malley, Pat, 179
On-demand economy, 180–186, 203
1 percent, the, 170–171, 190
Open content creation, 155
Open source, 31, 48, 153, 200
Organizing labor, 19, 89, 147–157
Orwell, George, 78
Oxfam, 215

Packard, Vance, 198
Panopticon, 8
Paolacci, Gabriele, 113
Paradox of participation, 156–157, 172, 203, 218
Pargman, Daniel, 202
Paris riots in 1968, 68
PARO, 37, 131–140, 143, 145–146, 174, 176. *See also* Social robots
Patterson, Don, 202
Paul, Chris, 123
PayPal, 122
Peer production and distribution, 48, 155, 156, 173, 175, 199, 201, 208
Penny, Simon, 13
Peppet, Scott, 179
Pepsi Co., 75
Personalization of risk, 74, 84–85, 120, 126, 157, 178–179, 198, 221
Petcou, Constantin, 190
Peters, Tom, 148
Petrescu, Doina, 190
Peterson, V. Spike, 46
PewDiePie, 102–103, 106, 171–172
Pierce, James, 198
Piketty, Thomas, 187, 190, 200, 207, 208, 221

Pinch, Trevor, 11
Pinker, Steven, 183
Planned obsolescence, 14, 198
Podolny, Shelley, 37
Polonyi, Michael, 41
Pope 100, 221
Postigo, Hector, 112, 118, 121, 125–126
Poverty, 43, 144–145, 154, 191, 195, 208
Power law, 152–156, 169–172, 190, 198, 206, 212, 215
 power law distribution, 153, 155, 156, 169, 172, 215
Precarity, 18, 32, 81, 84–86, 95, 99, 116, 123–127, 138–139, 168, 223
Predicaments, 4–7, 9, 18, 52–53, 64, 81–82, 84–88, 95, 102, 106, 124, 153, 156, 157, 161, 166, 169, 174, 177, 203, 216, 223
Presley, Elvis, 68, 172
Privatization, 74, 182
Proffitt, Jennifer, 13, 95
Punishment, 161, 164, 188

Raby, Fiona, 13
Racism, 68, 84
Raghavan, Barath, 197, 200
Rational choice, 45. *See also* Market
Reddit, 99
Reich, Robert, 184, 198, 221
Reitzle, Matthias, 154
Remy, Christian, 14
Repression, 52
ResearchGate, 100
Reuters, 167
Reynolds, Malvina, 68
Rheingold, Howard, 29
Ricardo, David, 42
Rifkin, Jeremy, 48, 81, 205–206
Riot Games, 111–113. *See also* League of Legends
Ritzer, George, 12, 35, 111
Rivera, Alex, 25
Robinson, James, 12, 62
Robinson, Marilynne, 129

Robotics, 12, 25, 110, 129–140, 143, 145–146, 174, 180–183, 188, 196
  drones, 55
  industrial robots, 180–183, 188
  Roomba and Neato, 130
  social robots, 19, 37, 129–140, 143, 145–146, 174
Rolling Stones, 71
Rooney, Ben, 182
Roos, Inger, 110, 148
Rose, Nikolas, 179
Rosendal, Henk, 130
Roszak, Theodore, 72
Rousseau, Jean-Jacques, 69
Rowling, J. K., 171
Russell, Steve, 117

Šabanović, Selma, 97, 129–130
Sawyer, R. Keith, 199
Schiano, Diane, 29
Schiller, Dan, 11, 51, 54, 65, 81, 180
Schmidt, Eric, 156, 170–171, 220
Schmidt, Florian Alexander, 121, 124, 126
Scholler, Dan, 19, 126, 129, 140–142
Scholz, Trebor, 12, 49
Schooler, Jonathan, 87
Schumacher, E.F., 73
Science and Technology Studies (STS), 11
Search Engine Optimization (SEO), 149
Seeger, Pete, 68
Self-control, 147–157, 188, 198–199, 203
Self-expression, 81, 100–101, 117, 165, 166, 168–169, 172, 211
Self-service, 108–109, 114–116, 140–141, 143, 164, 174, 218
Sengers, Phoebe, 198
Senior, Tom, 112
Sennett, Richard, 11, 23, 66, 81, 83, 168
Serfs, 34, 62, 147. *See also* Feudalism
Sevignani, Sebastian, 13
Shaker, Doug, 77
Shapiro, Carl, 219

Sharing economy, 47, 182, 184–186, 188, 198–199, 205–206
Shibata, Takanori, 131–140, 145–146
Shilton, Katie, 98
Shirky, Clay, 48, 152–155
Siegel, Jane, 83
Silbereisen, Rainer, 154
Silberman, M. Six, 60, 113–114, 144, 180, 198
Silicon Valley, 30–31
Simon, Herbert, 25, 27, 36, 212, 222
Singularity, Machine, 15
Slavery, 14, 34, 61, 65
Smith, Adam, 42, 43, 53, 184
Smythe, Dallas, 49
Snider, Laureen, 84
Snyder, Gary, 68
Social data, 96–106
Social equality, 46, 143–146, 154, 156, 185, 190, 198, 206, 214
Social media, 13, 32, 35, 38, 81, 83, 93–95, 98–106, 138, 142, 164, 174
Social safety net, 9, 32, 136, 154, 177, 182, 190, 194–195, 199, 212
Sood, Sara Owsley, 84
Sotamaa, Olli, 118
Spinoza, Baruch, 50
Sproull, Lee, 84
Sprouse, Jon, 113
Standing, Guy, 85
Starbucks Coffee, 221
Start-ups, 29, 77
Stiglitz, Joseph, 190, 221
Suarez-Villa, Luis, 11, 13, 23, 74, 84, 190
Subsistence, 190–201, 206–209, 217. *See also* Sustainability
Suchman, Lucy, 11
Suler, John, 84
Summers, Larry, 215
Sung, Ja-Young, 130
Sun Microsystems, 28, 76–77
Surveillance, 65, 73, 76–79, 212
Sustainability, 14, 190–201, 206–209, 221. *See also* Subsistence

Index

Takeuchi, Yuichiro, 197
Tanaka, Fumihide, 130
Tanenbaum, Josh, 198
Tanenbaum, Karen, 198
Targett, Sean, 121
TaskRabbit, 182, 184
Taylor, T. L., 118
Technocapitalism, 13
Technological determinism, 16, 32–33
Terranova, Tiziana, 12, 49, 176
Texas Instruments, 183
TextBroker, 113
The Who, 71
Thornton, Richard, 15
3D printing, 197, 200, 205
Thrift, Nigel, 101
Totalized stimulation, 37, 73–79, 87, 132, 166–167
Trans-Atlantic Investment and Trade Partnership (TTIP), 201
Trans-Pacific Partnership (TPP), 201
TripAdvisor, 115
Tsui, Katherine, 130
Turkle, Sherry, 31, 136
Twain, Mark, 83
Twitch.tv, 174
Twitter, 93–94, 99, 104, 152, 168, 172, 212, 218

Uber, 182, 184, 198
Ueno, Naoki, 199
Uncompensated labor, 24, 48–49, 55, 59, 61, 82, 95, 111, 120, 127, 147, 166, 204, 205
United Airlines, 97
U.S. Department of Defense, 205
Utopia, 19, 187–209
  real utopia, 187, 196, 209

Value, 15, 17–18, 31–43, 49–55, 57–63, 106–116, 117–127, 176, 220
  chain, 54
  creation or generation of, 15, 18, 49–50, 55, 56, 82, 89, 106–116, 117–127, 147, 149, 172, 185, 220

extraction of, 17, 33, 39, 43, 55, 57, 59, 62–63, 96, 106, 107–116, 117–127, 136–138, 142–143, 149, 176, 186
  securing, 34, 43, 55, 95 (see also Hidden value)
  surplus value, 35, 43–45, 49, 55, 57, 59, 62, 203
Vandermeer, John, 197, 221
Van Dijk, José, 12, 35, 96–97, 100, 126
Varian, Hal, 219
Vertesi, Janet, 131
Video games, 31–32, 35, 107, 111, 117, 121–127, 167, 171–173, 176, 212, 218
Virilio, Paul, 79
VisiCalc, 27–28
Volkoff, Serge, 30
Vygotsky, Lev, 38, 143

Wada, Kazuyoshi, 130–131, 133, 135–140
Wagner, Peter, 14, 166
Wakkary, Ron, 198
Wales, Jimmy, 154
Wallerstein, Immanuel, 54
Wall Street, 70
Walmart, 221
Want, Roy, 76
Waterman, Anthony Michael Charles, 45
Web 2.0, 95–106, 152, 218
Webb, Cynthia, 97
Weber, Max, 11, 41, 44, 66, 211, 215
Weeks, Kathi, 185–186, 188, 203
Weizenbaum, Joseph, 136
Welfare, 9, 60, 72–74, 84, 177–178, 217
Wikipedia, 47, 59–60, 152–154
  wikinomics, 152
Williamson, Oliver, 46 48
Wilson, Timothy, 87
Winner, Langdon, 16
Woelfer, Jill, 18
Womack, James, 148
World Bank, 201

World Economic Forum, 97, 201
World War II, 24, 26, 63, 67–68, 73, 172, 199, 200, 204, 208
World of Warcraft, 2, 119–122, 125, 167
World Wide Web, 30–31
Wozniak, Steve, 13, 27–28
Wright, Erik Olin, 23, 44, 54, 81, 187, 217–219
Wright, Will, 126

Xerox, 29, 76–77

Yelp, 79, 104, 173
Young, Neil, 71
Yousefzai, Malala, 100
YouTube, 2, 17, 18, 93, 100, 102–106, 127, 171, 173, 174, 218

Zhou, Yongming, 169
Zittrain, Jonathan, 30
Zuboff, Shoshana, 58
Zuckerberg, Mark, 156, 170–171